THE ARCHAEOLOGY, GEOLOGY, AND PALEOBIOLOGY of STANTON'S CAVE

Grand Canyon National Park Arizona

Robert C. Euler, Editor

With Contributions By

Donald P. Elston
Robert C. Euler
C. W. Ferguson
Lyndon Lane Hargrave
C. R. Harington
Richard Hereford
Richard H. Hevly
Austin Long
Paul S. Martin
Robert R. Miller
John W. Olsen
Stanley J. Olsen
Amadeo M. Rea
Eleanora I. Robbins
Gerald R. Smith

1984

Published by the
Grand Canyon Natural History Association
Monograph Number 6

ISBN 0-938216-21-X
Library of Congress Catalog Card Number: 84-80572
Editorial: Sandra Scott, Margaret Stephan
Production: Sandra Scott, Margaret Stephan, John C. O'Brien
Lithography: Classic Printers; Prescott, Arizona
Cover Illustration: Robert C. Euler

Acknowledgments

I should like to express my sincere appreciation, first, to the National Geographic Society for two grants-in-aid that made the field work in Stanton's Cave possible.

Archaeologists who accompanied me on the first, month-long excavation in the cave in 1969 were Bruce Harrill, Lawrence Powers, Robert Page, and John A. Ware III. The latter two were undergraduate students of mine at Prescott College; Powers had been my student previously at Northern Arizona University; Harrill was then a graduate student at the University of Arizona. All of them worked very hard in the field, under adverse conditions with temperatures often climbing above 100°F, and were all good companions during the month we were isolated in the depths of Marble Canyon. I am deeply indebted to them all.

A special note of thanks must go to Wayne Learn, our helicopter pilot, who successfully flew all personnel and equipment into the canyon on both field trips and who brought specialists and fresh food in to our camp from time to time. Our safety depended upon his skill as a helicopter pilot and he and his Hiller helicopter served us very well indeed.

On the second field trip I was aided by Paul Long, photographer; Steven Clarke and Deborah Westfall, at that time my archaeology students at Prescott College; Bruce Harrill, Barney Burns and Martha Ames, graduate students in anthropology and geosciences at the University of Arizona; Paul S. Martin and Austin Long, on the faculty of the Department of Geosciences at the University of Arizona; and C. W. Ferguson, of the faculty of the Laboratory of Tree-Ring Research at the same university.

Other specialists who consulted with and aided us in the field on our first trip were Paul S. Martin; Walter W. Taylor, then an archaeologist at Southern Illinois University, Carbondale; David Schleicher, a geologist with the U.S. Geological Survey; and Stanley J. Olsen, a paleontologist at the University of Arizona.

One of the most valuable contributions made in the field by these persons, in addition to assisting us with our excavations, was their participation, along with my students, in impromptu but exciting seminars held at our field camp on the beach along the Colorado River. This, in my mind, represented some of the best of scientific education for the students, and I am most grateful to all who participated.

Subsequent to our two excavation trips, I have visited the cave on two occasions to obtain samples for paleomagnetic dating. These were facilitated by the National Park Service, the U.S. Geological Survey, and the volunteered efforts of the following Survey personnel: Henry Holt, Don Elston, Joe Rosenbaum, Chuck Naeser, Hugh Reich, Jim Scott, and Gary Olhoeft; my thanks to all of them for their work in the dusty confines of Stanton's Cave.

Appreciation is extended to Steve Robinson of the U.S. Geological Survey's radiocarbon laboratory, Menlo Park, California, for the most recent (1984) ^{14}C date of Stanton's Cave driftwood.

I also very much appreciate the efforts of all the authors of the reports included in this volume. They have been most patient in the face of long delays and sometimes seemingly insurmountable obstacles in getting this report published.

Thanks are also due the Grand Canyon Natural History Association and its board of directors for establishing a research monograph series and for including Stanton's Cave report in it.

Federal permits for the excavation were issued by the Department of Interior and Grand Canyon National Park on April 17, 1969.

R.C.E.

Table of Contents

Chapter

1. Introduction 1
Robert C. Euler
2. The Archaeology and Geology of Stanton's Cave 7
Robert C. Euler
3. Macroscopic Plant Materials 33
Richard H. Hevly
4. Zooarchaeological Analysis of Small Vertebrates 47
John W. Olsen & Stanley J. Olsen
5. The Fish Remains, with Notes on the Taxonomy of *Gila cypha* 59
Robert R. Miller & Gerald R. Smith
6. The Ungulate Remains: An Identification List 67
C. R. Harington
7. The Bird Bones 77
Amadeo M. Rea & Lyndon Lane Hargrave
8. Dendrochronology of Driftwood from Stanton's Cave 93
C. W. Ferguson
9. Driftwood in Stanton's Cave: The Case for Temporary Damming of the Colorado River at Nankoweap Creek in Marble Canyon 99
Richard Hereford
10. Polarity of River-Flood Silt in Stanton's Cave 107
Donald P. Elston
11. Paleoecology of Stanton's Cave 115
Eleanora I. Robbins, Paul S. Martin & Austin Long
12. Stanton's Cave During and After the Last Ice Age 131
Paul S. Martin
13. Conclusions 139
Robert C. Euler

Appendix 2-A Description of figurines recovered under controlled conditions in 1963, 1969, and 1970. 19
Appendix 2-B Miscellaneous Stanton's Cave figurines. 28
Appendix 3-A List of macroscopic plant materials from Stanton's Cave. 43
Appendix 3-B Cuticle identifications from artiodactyl scats from Stanton's Cave. 46

List of Illustrations

Figure Number

1-1. Stanton's Cave in Marble Canyon, between Vaseys Paradise and the mouth of South Canyon.
1-2. Stanton's Cave (arrow), deep in the Redwall Limestone of Marble Canyon.
1-3. The mouth of Stanton's Cave.
1-4. Vaseys Paradise, a few hundred meters below Stanton's Cave.
1-5. All personnel, equipment, and supplies were flown to the site by helicopter.
1-6. The expedition camp on the beach at the mouth of South Canyon.
1-7. Prescott College — National Geographic Society Stanton's Cave Archaeological Expedition Personnel. June 12-July 11, 1969.
1-8. Prescott College — National Geographic Society Stanton's Cave Archaeological Expedition Personnel. September 20-27, 1970.
2-1. Split-twig figurines from Stanton's Cave are of generally similar construction.
2-2. Ground plan and cross sections, Stanton's Cave.
2-3. The rock-strewn floor of the main room of the cave before excavation.
2-4. Antechamber 1, behind the main room of Stanton's Cave.
2-5. Profiles of the two test trenches in Stanton's Cave.
2-6. North-South Trench looking north.
2-7. Closeup view of figurine in Figure 2-6.
2-8. North-South Trench in main room of Stanton's Cave looking south.
2-9. Grids O-P-Q-19 in North-South Trench.
2-10. West wall of grids O-19 and P-19, North-South Trench.
2-11. East-West Trench looking west.
2-12. Grids GG and HH in East-West Trench.
2-13. South wall of grids EE and FF in East-West Trench.
2-14. Vertical view of Grid I-I to the 60 cm level.
2-15. Vertical view of Grid I-I to the 1 meter level.
2-16. Cache of 3 figurines (F.S. #17) *in situ.*
2-17. Cache of 4 figurines (F.S. #19) *in situ.*
2-18. Cache of 3 figurines (F.S. #24) *in situ.*
2-19. Two caches of figurines (F.S. #25 left and F.S. #26 right) *in situ.*
2-20. Single figurine (F.S. #30) probably *in situ.*
2-21. Figurines *in situ,* F.S. #34 on right and F.S. #35 on left.
2-22. Yucca cordage.
2-A-1. Catalog number C:5:3.1.
2-A-2. Catalog number C:5:3.2.
2-A-3. Catalog number C:5:3.3.
2-A-4. Catalog number C:5:3.4.
2-A-5. Catalog number C:5:3.5.
2-A-6. Catalog number C:5:3.6.
2-A-7. Catalog number C:5:3.7.
2-A-8. Catalog number C:5:3.8.
2-A-9. Catalog number C:5:3.9.
2-A-10. Catalog number C:5:3.10.
2-A-11. Catalog numbers .11, .13 and .14.
2-A-12. Catalog numbers .15, .17 and .18.
2-A-13. Catalog numbers .21, .22, .24 and .25.
2-A-14. Catalog numbers .26 and .27.
2-A-15. Catalog numbers .28, .29 and .30.
2-A-16. Catalog numbers .31, .32 and .33.
2-A-17. Catalog numbers .34, .35, .36 and .39.
2-A-18. Catalog numbers .40 and .54.
2-A-19. Catalog numbers .41 and .42.
2-A-20. Catalog numbers .43 and .47.
2-A-21. Catalog number C:5:3.48.
2-A-22. Catalog numbers .50 and .51.
2-A-23. Catalog numbers .55 and .56.
2-A-24. Catalog numbers .58, .59., .60 and .61.
2-A-25. Catalog numbers .63, .66, .67 and .69.
2-A-26. Catalog numbers .70, .71 and .72.
2-A-27. Catalog number C:5:3.73.
2-A-28. Catalog number C:5:3.92.
2-B-1. Miscellaneous Stanton's Cave figurines.
3-1. Seed composition of sediments from Stanton's Cave.
3-2. Percentages of pollen types recovered from Grid I-I compared with modern pollen samples obtained outside Stanton's Cave.
3-3. Concentration and cuticle analysis of artiodactyl scats from Stanton's Cave.
5-1A. Lateral view of neurocranium of *Gila robusta.*
5-1B. Lateral view of neurocranium of *Gila elegans.*
5-1C. Lateral view of neurocranium of *Gila cypha.*
5-1D. Incomplete neurocranium of *Gila cypha* from Stanton's Cave.
5-2A. Posterior 14 caudal vertebrae of *Gila robusta.*
5-2B. Posterior 14 caudal vertebrae of *Gila cypha.*
5-2C. Posterior 13 caudal vertebrae of *Gila elegans.*
5-2D. Nine caudal vertebrae of *Gila elegans* from Stanton's Cave.
5-3A. Dorsal view of left pectoral girdle of *Gila elegans.*
5-3B. Dorsal view of left pectoral girdle of *Gila cypha* from Stanton's Cave.
5-3C. Incomplete neurocrocranium of *Catostomus latipinnis.*
5-3D. Incomplete left dentary of *Catostomus discobolus.*
7-1. Lateral view of robin crania.
7-2. Palatal view of robin skulls.
7-3. Turkey and robin distribution.
8-1. Cross section of Douglas-fir log (STC-1).
8-2. Cross section of Douglas-fir log (STC-2).
8-3. Histogram of distribution of measured tree-ring series.
9-1A. Marble Canyon and Stanton's Cave in the eastern Grand Canyon, Arizona.
9-1B. Topographic map of the Nankoweap area.
9-2. Surficial geologic map of the Nankoweap Rapids area.

9-3. Exposure of the older Pleistocene colluvium near elevation 890 m (Figure 9-2).
9-4. Small clasts and matrix of the older colluvium.
9-5. Generalized cross section of Marble Canyon at Nankoweap Rapids.
10-1. Generalized sections of trench walls in Stanton's Cave.
10-2. Equal area plots of declination and inclination of 30 oriented samples of silt from Stanton's Cave.
10-3. Equal area plot showing directions resulting from progressive alternating field (AF) demagnetization in the range of 50-950 oersteds.
10-4. Orthogonal demagnetization diagram showing projection of vector end point during progressive AF demagnetization analysis.
11-1. Location of Stanton's and Rampart caves, Grand Canyon, Arizona.
11-2. Stanton's Cave and Vaseys Paradise.
11-3. Modern and fossil artiodactyl fecal pellets.
11-4. Top: weight distribution of fecal pellets of bighorn sheep, Harrington's mountain goat, mule deer, and extant mountain goat.
11-5. Graphic analysis and comparison of weight, length, and width of fecal pellets.
11-6. Arithmetic plot of radiocarbon dates versus depths in Stanton's Cave.
11-7. Photomicrographs of fossil epidermal fragments in pellets.
11-8. Pollen profile of cave earth, Stanton's Cave.
12-1. Aerial view of Stanton's Cave.
12-2. Profile of wall of Grid I-I.
12-3. North-South Trench in main room of Stanton's Cave.

List of Tables

Table Number	
2-B-1.	Catalog numbers of miscellaneous figurines from Stanton's Cave.
3-1.	Macroscopic biotic remains from Stanton's Cave.
4-1.	Taxonomic list of small mammalian fauna present.
4-2.	Mammalian fauna: quantitative data.
4-3.	Occurrence of mammalian remains by horizontal provenance.
4-4.	Taxonomic list of herpetofauna present.
5-1.	Fish remains from Stanton's Cave.
6-1.	Radiocarbon dates of horn sheaths of *Oreamnos harringtoni.*
7-1.	Occurrences of bird bones in Stanton's Cave.
8-1.	Proveniences of Stanton's Cave Douglas-fir driftwood.
11-1.	Plants identified in the fossil fecal pellets, Stanton's Cave.
11-2.	Summary of data from fecal pellets removed from Stanton's Cave.
11-3.	Radiocarbon dates from Stanton's Cave.

Institutional Affiliations of the Authors

Elston, Donald P.	U.S. Geological Survey, Flagstaff, Arizona
Euler, Robert C.	Grand Canyon National Park and Arizona State University, Tempe
Ferguson, C. W.	The University of Arizona, Tucson
Hargrave, Lyndon Lane	Deceased, formerly of Prescott College, Arizona
Harington, C. R.	The National Museums of Canada, Ottawa
Hereford, Richard	U.S. Geological Survey, Flagstaff, Arizona
Hevly, Richard H.	Northern Arizona University, Flagstaff
Long, Austin	The University of Arizona, Tucson
Martin, Paul S.	The University of Arizona, Tucson
Miller, Robert R.	The University of Michigan, Ann Arbor
Olsen, John W.	The University of Arizona, Tucson
Olsen, Stanley J.	The University of Arizona, Tucson
Rea, Amadeo M.	San Diego Natural History Museum
Robbins, Eleanora I.	U.S. Geological Survey, Reston, Virginia
Smith, Gerald R.	The University of Michigan, Ann Arbor

Chapter 1

Introduction

by

Robert C. Euler
Grand Canyon National Park
and
Department of Anthropology
Arizona State University, Tempe

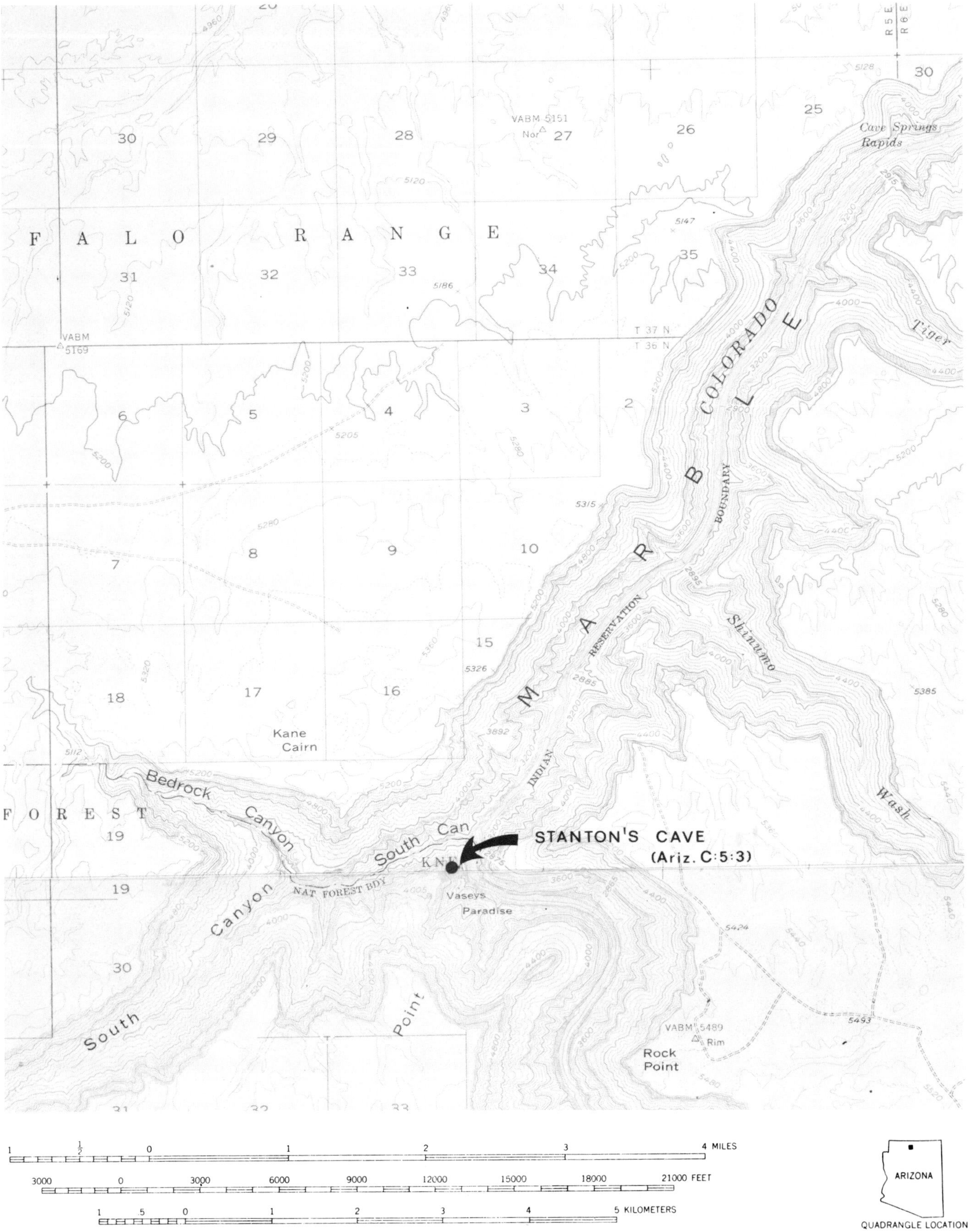

Figure 1-1. Stanton's Cave in Marble Canyon, between Vaseys Paradise and the mouth of South Canyon.

Stanton's Cave is a large solution cavern in the Mississippian Redwall formation in the inner gorge of Marble Canyon, an easterly extension of Grand Canyon (Figure 1-1). The cave is 31.7 miles (51 river kilometers) below Lees Ferry[1], Arizona, and is at an elevation of 3041 feet (927 meters) (Figure 1-2). The cave itself is 144 feet (44 meters) above the present, post-Glen Canyon Dam, level of the Colorado River. It is recorded as archaeological site Ariz. C:5:3 in the Grand Canyon archaeological survey.

In front of the cave, the talus, composed mostly of large blocks of Redwall limestone, slopes steeply at an angle of 32° to the rocky shore of the river (Figure 1-3). The level of the river fluctuates a few meters each day due to demands for hydroelectric power at Glen Canyon Dam upstream.[2]

Vegetation around the cave is scanty, except at a large spring, Vaseys Paradise, about 200 meters downstream from the cavern (Figure 1-4). Aside from the exotic tamarisk *(Tamarix chinensis)*, the principal plants outside the cave are agave *(Agave utahensis)*, Mormon tea *(Ephedra* sp.), beavertail cactus *(Opuntia basilaris)*, barrel cactus *(Echinocactus polycephalus)*, buffaloberry *(Sheperdia argentea)*, sparse small grasses, and, along the river, willow *(Salix* sp.).

Figure 1-2. Stanton's Cave (arrow), deep in the Redwall Limestone of Marble Canyon.

Figure 1-3. The mouth of Stanton's Cave, 144 feet (44 meters) above the Colorado River.

Figure 1-4. Vaseys Paradise, a few hundred meters below Stanton's Cave.

Animals are not abundant; the only land species observed included bats, lizards, rock squirrels, and snakes. Predominant birds noted were cliff swallows, violet green swallows, canyon wrens, hummingbirds, and ravens.

Foot access to Stanton's Cave is restricted to two routes: across the river at kilometer 48.6 from the left bank where the canyon walls are broken and faulted and then downstream to the cave; and an easier route down South Canyon, a right bank tributary that enters the Colorado a few hundred meters upstream from the cave.

Stanton's Cave is named for the intrepid engineer, Robert Brewster Stanton, who in 1889-1890 surveyed for a water-level railroad route through Grand Canyon. In July of 1889, after three men in Stanton's party had drowned within five days, he cached his supplies in the cave and hiked out South Canyon, ultimately reaching the Mormon settlement at Kanab, Utah. Parenthetically, within six months, Stanton was back on the river, this time equipped with life jackets, and, incredible as it may sound, successfully completed his survey. There is no indication in Stanton's journal that he observed any cultural or other remains within the cave, although he did note a small prehistoric Anasazi (or Pueblo) ruin at the mouth of South Canyon as he and his men began their laborious hike out of the gorge (Smith 1965:89). This pueblo site (Ariz. C:5:1 in the Grand Canyon archaeological survey) is still observable and consists of several isolated single rooms along a limestone shelf some 10 to 20 meters above the river. It is a Kayenta Anasazi ruin occupied seasonally *ca.* A.D. 1050 to 1150.

It was not until 1934[3] that prehistoric archaeological remains were found in Stanton's Cave. Reilly (1966;130-131, 134) quotes from the diary of Alton Hatch, a river runner in that year, to the point that "Bus Hatch found several little horses made out of willows and sticks." These were in fact split willow twig figurines representative of some type of large four-legged animal, perhaps deer or bighorn sheep. In all probability, their manufacture involved some form of imitative magic since some of them seem to have been ritually pierced with tiny wooden "spears." Perhaps their makers imitated the successes they hoped for in their actual hunts and carefully placed the effigies in caches in isolated caves such as Stanton's.

In 1963 a figurine that I recovered *in situ* in Stanton's Cave was radiocarbon dated at 4095 ± 100 years before the present (B.P.) (UCLA - 741A; Berger, Fergusson and Libby 1965:340). This date agrees well with radiocarbon age estimates for similar figurines recovered from other caves in Arizona and Nevada.

While no *diagnostic* cultural remains have been found in *direct association* with the figurines, it has been postulated that they may have been produced by hunters of the Archaic Pinto Basin (Euler and Olson 1965; McNutt and Euler 1966; Euler 1966) or Gypsum (Schroedl 1977) complexes. At present, the finds are referred to archaeologically as the Grand Canyon Figurine Complex. This will be discussed in the next chapter.

In recent years, prior to the excavations reported here, Stanton's Cave had been visited by thousands of people on Colorado River raft trips as well as by hikers coming down South Canyon and by amateur cave explorers. The site was being increasingly disturbed, especially by the latter (Anonymous 1966). Estimates of the total number of figurines taken illegally[4] from the cave between 1934 and our controlled excavations number more than 165.

In addition to the presence of figurines in Stanton's Cave, amateur explorers had also reported what later proved to be the remains of Pleistocene fauna in the cavern (Parmalee 1969).

The scientific importance of the site, therefore, was twofold: it held promise of yielding more details as to the cultural affinities of the split-twig figurine makers and of providing data regarding the paleoenvironment of Grand Canyon from Pleistocene times to the present.

In order to prevent further vandalism and to recover scientific data, the National Geographic Society provided funds for the construction of a protective fence at the mouth of the cave[5] and, in 1969 and 1970, made grants to the author to conduct archaeological, paleontological, and biological excavations there.

The research design for the project set forth two major goals. The first was to attempt to locate additional figurines, undisturbed *in situ, in direct association with other diagnostics,* to test the validity of the hypothesis that the figurines were made by Pinto Basin hunters. The second was to conduct the excavations with such care stratigraphically that a total ecologically oriented recovery could be made of the cultural and biological remains in a datable context. This, of course, was aimed at reconstructing the paleoenvironment of the cave environs and those of Grand Canyon in general over whatever time span was represented by the deposits.

Figure 1-5. All personnel, equipment, and supplies were flown to the site by helicopter.

During all field seasons, personnel and excavation and camp equipment were taken to the site by helicopter (Figure 1-5). In addition, the USGS party to take paleomagnetic samples in 1982 arrived by boat. A camp was established on a sand bar along the river a few hundred meters upstream from the cave (Figure 1-6). The 1969 excavations continued for 30 days during June and July. In 1970, the party was in the field for seven days in September. The initial recovery of samples for paleomagnetic dating took place for five days in November 1976 at the same time geological mapping of a hypothetical landslide at the mouth of Nankoweap Canyon, 32.7 kilometers downstream, was carried out (see Hereford, this volume). The most recent collection of paleomagnetic samples occurred during a two-day period in May 1982 (see Elston, this volume).

The results of the excavations and analyses constitute the remaining chapters of this book. A report of progress of the study has been published earlier (Euler 1978).

Figure 1-6. The expedition camp on the beach at the mouth of South Canyon.

Figure 1-7. Prescott College — National Geographic Society Stanton's Cave Archaeological Expedition Personnel. June 12-July 11, 1969. Left to right: Bob Page, Larry Powers, John Ware, Bruce Harrill, Bob Euler.

Figure 1-8. Prescott College — National Geographic Society Stanton's Cave Archaeological Expedition Personnel. September 20-27, 1970. Left to right: Austin Long, Bob Euler, Debbie Westfall, Wes Ferguson. Middle: Steve Clarke, Martha Ames, Bruce Harrill, Barney Burns. Rear: Paul Martin.

Footnotes

1. Figures given in Euler 1978:141 were in error. The cave's UTM coordinates are 12/423125 m E and 4039600 m N.
2. In June 1983 the Bureau of Reclamation found it necessary to release as much as 92,000 cubic feet per second from the dam a brief period, thus temporarily altering this regime.
3. The date of 1939 given in Euler 1978:142 and Euler 1966:62 was in error. See Reilly 1966:130-131 and 134 for data on the 1934 Hatch river trip when figurines were first discovered in Stanton's Cave.
4. That is, without federal antiquities permits. One group of these, recovered by the National Park Service, is described and illustrated in Appendix 2-B.

5. Unfortunately, the fence and locked gate continue to be breached by vandals and curiosity seekers.

References

Anonymous
1966 Split-twig figurines found by CAG. *Cave Crawlers Gazetteer* 7:4.

Berger, Rainer, G. J. Fergusson, and W. F. Libby
1965 UCLA Radiocarbon Dates IV. *Radiocarbon* 7: 340-341.

Euler, Robert C.
1966 Willow Figurines from Arizona. *Natural History* 75:3:62-67.
1978 Archeological and Paleobiological Studies at Stanton's Cave, Grand Canyon National Park, Arizona — A Report of Progress. *National Geographic Society Research Reports, 1969 Projects* pp. 141-162.

Euler, Robert C. and Alan P. Olson
1965 Split-twig Figurines from Northern Arizona: New Radiocarbon Dates. *Science* 148:368-369.

McNutt, Charles H. and Robert C. Euler
1966 The Red Butte Lithic Sites Near Grand Canyon, Arizona. *American Antiquity* 31:3:410-419.

Parmalee, Paul W.
1969 California Condor and Other Birds from Stanton Cave, Arizona. *Journal of the Arizona Academy of Science* 5:4:204-206.

Reilly, P. T.
1966 The Sites at Vasey's Paradise. *The Masterkey* 40:4:126-139.

Schroedl, Alan R.
1977 The Grand Canyon Figurine Complex. *American Antiquity* 42:2:254-265.

Smith, Dwight L. (ed.)
1965 *Robert Brewster Stanton: Down the Colorado.* Norman: University of Oklahoma Press, 237 pp.

Chapter 2

The Archaeology and Geology of Stanton's Cave

by

Robert C. Euler

Grand Canyon National Park

and

Department of Anthropology
Arizona State University, Tempe

The first professional archaeological reconnaissance in Stanton's Cave took place in 1963. The author, in the company of the late Alan P. Olson, recovered 10 complete and 10 fragmentary split-twig figurines during a brief visit. These artifacts were located in three separate caches under rock fall inside the left portion of the mouth of the cave (there was insufficient time during this initial visit to draw a ground plan).

The range in variation of the complete specimens was from 7.7 to 19.0 cm long and 9.2 to 14.5 cm high, an average of 11.8 cm long and 11.3 cm high. Seven of the ten figurines were pierced obliquely through the body with split twigs ranging in length from 8.9 to 31.0 cm, an average of 19.0 cm. Body wraps ranged in number from 3 to 5, averaging 4.3, while neck wraps ranged from 3 to 7, averaging 4.8.[1]

One fragmentary figurine from this collection was dated by the Radiocarbon Laboratory of the University of California at Los Angeles (UCLA-741) at 4095 ± 100 years B.P. (2145 ± 100 B.C.) (Euler and Olson 1965). This is the oldest date yet determined for the Grand Canyon Figurine Complex.

The figurines recovered from our initial investigations conform generally to other figurines previously recorded from the Grand Canyon (Euler 1966). These have been described in sufficient detail in other papers (Farmer and deSaussure 1955; Schwartz, Lange, and deSaussure 1958) that a brief summary will suffice here. They were simply but ingeniously made from a branch of willow *(Salix* sp.), or occasionally of cottonwood *(Populus fremontii)* or squawbush *(Rhus trilobata)* (Bohrer 1983). This branch was split longitudinally down most of its length. The unsplit portion was bent to form the rear legs and back; one splint was then bent to form the front leg, neck, and head, while the other was wrapped around both legs to form the body (Figure 2.1). Occasionally, the body area contains a fecal pellet, probably from either deer *(Odocoileus hemionis)* or bighorn sheep *(Ovis canadensis)*. Some are pierced through the body with split twigs.

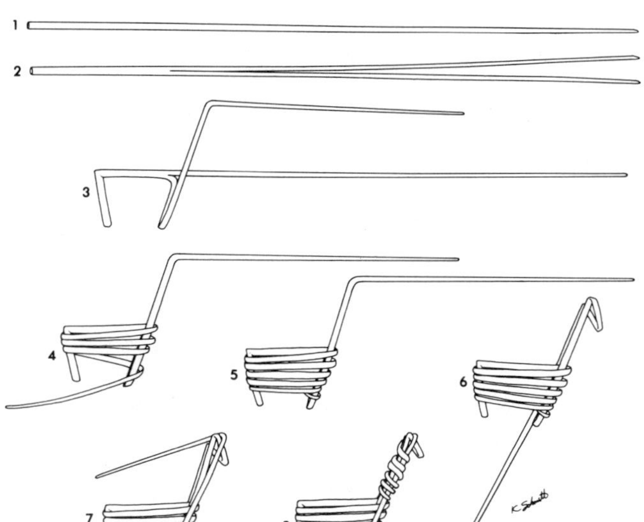

Figure 2-1. Split-twig figurines from Stanton's Cave are of generally similar construction, made from one twig (1) which was split to within several centimeters of its butt end (2). The unsplit portion formed the back leg and backbone (3). One splint, when wrapped around the legs (4) formed the body (5). The other splint which formed the front leg was then brought up to construct the head and neck (6-8). (Drawing by Karen Schmitt)

The range of distribution of these artifacts is from southern California on the Mohave Desert, southeastern Nevada, northern Arizona, and southeastern Utah although those from the latter region are more variable in construction with the addition of vertical wraps around the body (Schroedl 1977).

There are sufficient radiocarbon dates to indicate that the Grand Canyon figurines were manufactured between 3000 and 4000 years ago (see Schroedl 1977:256 for a summary table of locations of figurine discoveries and radiocarbon dates).

It is generally agreed that they represented some form of magico-religious ritual, the twigs inserted through the body having been representations of spears that functioned to insure success in the actual hunt.

I have earlier suggested that the makers of the figurines were part of the Pinto Basin Complex, since diagnostic Pinto points, dating from approximately the same age, have been found in a similar geographic range (Euler and Olson 1965; McNutt and Euler 1966; Euler 1966; Euler 1983). More recently the recovery of figurines somewhat resembling those from Grand Canyon from Cowboy Cave in southeastern Utah, has resulted in a suggestion that they may have been made by people of the Gypsum tradition (Schroedl 1977). A number of figurines in direct association with Gypsum points were recovered from levels radiocarbon dated at 1685 B.C. and 1610 B.C. (Jennings 1980:24). Figurines themselves were not dated.

While no Gypsum points, nor for that matter, any Archaic evidence other than Pinto have been found at Grand Canyon, several Pinto points have been recovered from the South Rim of Grand Canyon and from the summit and slopes of Red Butte, some 25 km south of Grand Canyon (McNutt and Euler 1966; Euler 1983). The hypothesis that people of the Pinto Complex were the makers of the Grand Canyon split-twig figurines remains viable. Excavations in Stanton's Cave were conducted, in part, to test this hypothesis.

Stanton's Cave is a very large grotto. Its naturally roofed entry is a steep, rocky talus, 21 meters long and varying in width from 3 to 10 meters (Figure 1-3). The main room of the cave is approximately 17 by 44 meters (Figure 2-2), with a fairly level rock-strewn floor (Figure 2-3). At the rear of the main room, the floor begins to slope sharply upward. There, a narrow crawlway provides access to a high vaulted rear antechamber (Figure 2-4) which was designated Antechamber 1.

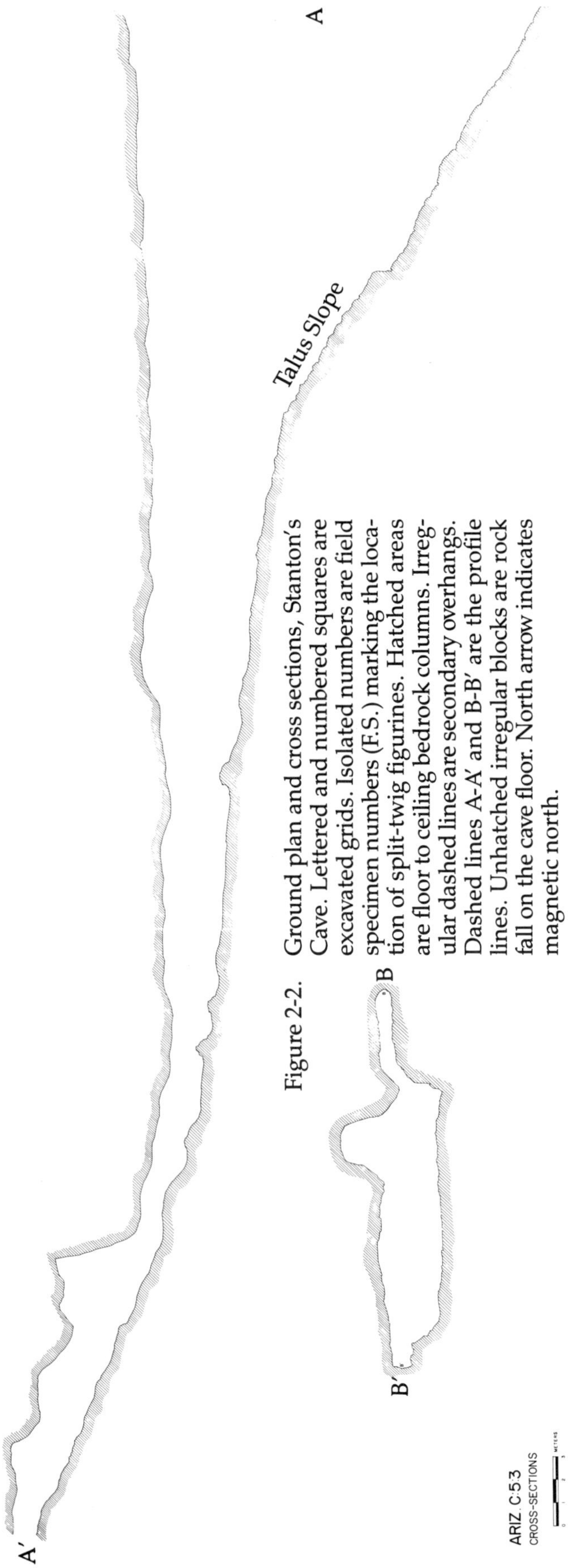

Figure 2-2. Ground plan and cross sections, Stanton's Cave. Lettered and numbered squares are excavated grids. Isolated numbers are field specimen numbers (F.S.) marking the location of split-twig figurines. Hatched areas are floor to ceiling bedrock columns. Irregular dashed lines are secondary overhangs. Dashed lines A-A′ and B-B′ are the profile lines. Unhatched irregular blocks are rock fall on the cave floor. North arrow indicates magnetic north.

The walls and ceiling of the cave, according to David Schleicher (personal communication), who studied the surficial geology of the cave during the project, occur in limestone that is crossed by joints a few centimeters wide filled with calcareous white clay, perhaps illite. No stalactites or stalagmites were observed.

At the outset of the 1969 field season, a map of the cave, both ground plan and cross sectional views (Figure 2-2), was made so that all artifacts and other specimens recovered could be accurately located in horizontal and vertical proveniences. A careful search was then made of the entire cave for surface artifacts. None was found in the crawlway or the rear antechamber although a number of biological specimens were recovered from pack rat nests in those reaches of the cave.

After clearing the rock fall from the floor of the main room, during which a number of figurines were recovered under the rocks, two grid systems were laid out. One grid, in a north-south line in the widest part of the room extended for 12 meters. A second grid, closer to the entrance and in an east-west direction, covered a length of 8 meters. Each was one meter wide. The north-south grids were marked F-19 through R-19, while the east-west squares were lettered AA through HH respectively.

Following that, the floor area that was free of rocks, especially out to 1 meter from the cave walls, was carefully trowelled to a depth of 10 cm. All figurines recovered from the entire excavation came from either the surface under the rock fall or from a depth of no greater than 10 cm in the dust of the cave floor. Each artifact or associated group of artifacts (since many of the figurines occurred in caches) was assigned a field specimen number (F.S.) and accurately located on the ground plan. These are shown with numbered arrows in Figure 2-2.

During the 1969 excavations, each square meter was excavated in arbitrary 5 cm levels from the present floor surface to a depth of 25 cm. In some cases, however, because of masses of rock in the East-West Trench, 25 cm levels had to be employed in preference to the tighter control. Below 25 cm, material was removed in arbitrary 25 cm levels until bedrock was reached at a maximum depth of 1.5 meters. All excavated material was screened, that from the 5 cm levels through a window screen and the other through a 6 mm (¼″) mesh.

Subsequent laboratory analyses indicated a need for even more refined vertical controls. Therefore, the 1970 season, in addition to accomplishing specialized biological tasks, included the excavation of another one meter square area (Grid I-I; originally referred to as Stratigraphic Test Pit #1), adjacent to the earlier east-west trench. The fill was removed in 5 cm levels from surface to bedrock and *all* cultural and biological specimens were collected after screening through a window screen. Random samples of materials that passed through that mesh were taken to the river for recovery of minute specimens by a flotation process. The aim was to make as total an ecological recovery as possible.

The present floor of the cave is underlain by unconsolidated silts, locally mantled by a thin layer of organic

Figure 2-2.

debris (Figure 2-5). Schleicher (personal communication) noted that in both trenches where bedrock was exposed, it consisted of non-lustrous, apparently degraded flowstone. Pale gray-brown silts and clayey silts covered the bedrock thickening toward the center for the main room.

In the North-South Trench, the bedrock was covered by about 50 cm of stony, pale brown silt. This silt contained about 25% limestone fragments typically a few tens of centimeters in dimension; these fragments were mostly heavily concentrated near the cave wall. Resting on the silt were entangled masses of driftwood. Relatively little fine debris was between the logs, but the interstices did contain numerous artiodactyl dung pellets. White and tan silt surrounded the logs, many of which were resting on bedrock, and these silts showed no apparent stratification or compaction. Included were occasional delicate helictites a few millimeters long. A tongue of tan clayey silt extended south in the trench into the white clay. Preliminary analyses indicate that the clay is dominantly illite (Figures 2-6 to 2-10).

In the East-West Trench, the stratigraphy was less regular and the bedrock extended toward the mouth of the cave in rough steps. Most of the fill consisted of pale gray-brown clayey silt about 50 to 80 cm thick. Two irregular and discontinuous layers between 10 and 20 cm

Figure 2-4. Antechamber 1, behind the main room of Stanton's Cave. Vertical scale is 1 meter.

Figure 2-3. The rock-strewn floor of the main room of the cave before excavation.

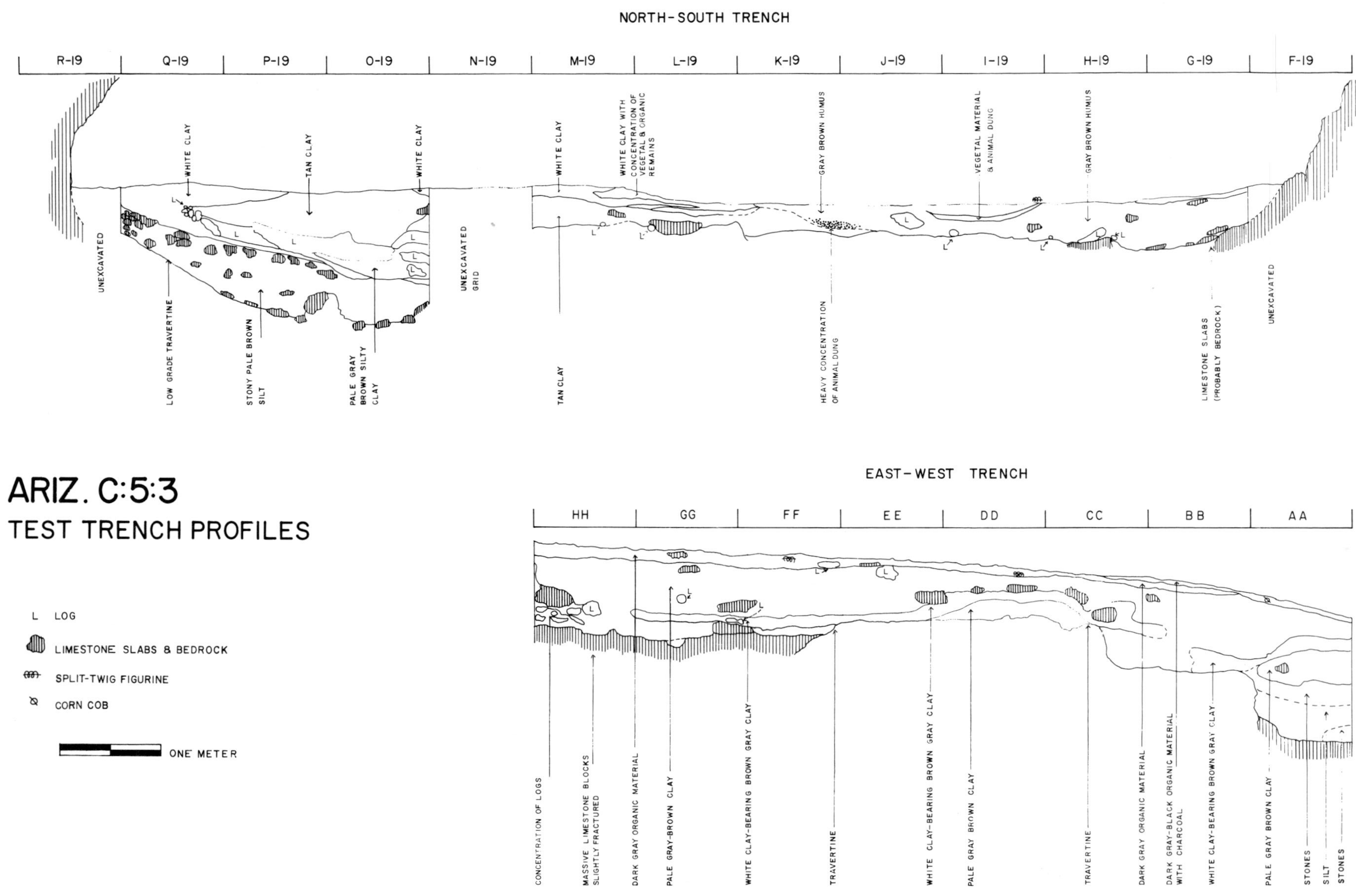

Figure 2-5. Profiles of the two test trenches in Stanton's Cave.

thick were relatively enriched in white clay; for part of the length they lay on bedrock, but in part were within the gray-brown clayey silt. Several logs also occurred within it, one resting nearly on bedrock and one on the upper surface of the clay. The clayey silt is covered by a mantle about 5 to 15 cm thick of dark gray organic material which makes up the present cave floor (Figures 2-11 to 2-13).

The 1 meter square labeled I-I yielded vegetal matter to a depth of 10 cm, with gray-tan powdery dirt and small rock spalls to 15 cm. Below that, larger rocks above driftwood were encountered until, at a depth of 60 cm, approximately 75% of the square was covered with rock and driftwood (Figure 2-14). Bedrock was reached at a depth of 1 meter (Figure 2-15).

Figure 2-6. North-South Trench looking north. Vertical scale is 1 meter; arrow points to figurine *in situ*.

The Figurines and Figurine Caches

As has been noted, all figurines from our controlled excavations were recovered under rock falls or cave floor dust at a depth not exceeding 10 cm. Some additional specimens were found in pack rat nests and hence lack a provenience. Seventy-four figurines, complete and fragmentary, were recovered. All artifacts are now curated at Grand Canyon National Park.

All figurines were made in accord with the general pattern that has been described above. The range in variation of the complete specimens was from 5.1 to 16.4 cm long and 7.1 to 19.5 cm high, an average of 9.1 cm long and 10.9 cm high. Only two of the figurines were pierced, spearlike, with a split twig. These were 27.5 cm and 13.4 cm long respectively. Two other figurines had hornlike appendages to the head. Body wraps ranged in number from 3 to 9, averaging 5.3, while neck wraps ranged from 2 to 10, averaging 5.[1]

Many of the figurines recovered *in situ* were in caches, groups of two or more, placed under rocks that had

Figure 2-7. Closeup view of figurine in Figure 2-6 (black arrow). Driftwood and artiodactyl dung in bottom of excavation. Grids I-19 and H-19, North-South Trench.

Figure 2-8. North-South Trench in main room of Stanton's Cave looking south. North arrow is 30 cm long; vertical scale graduated in 10 cm increments. Note driftwood.

fallen from the cave roof. Each cache or isolated figurine recovered was given a field specimen number (F.S. #) and the nature of these is described here.

F.S. #3[2] — A complete figurine (C:5:3.11) recovered from a pack rat midden; no other provenience.

F.S. #4 — A fragmentary figurine (C:5:3.12) wedged in rocks to rear of main room; probably *in situ*.

F.S. #5 — A complete figurine (C:5:3.13) recovered

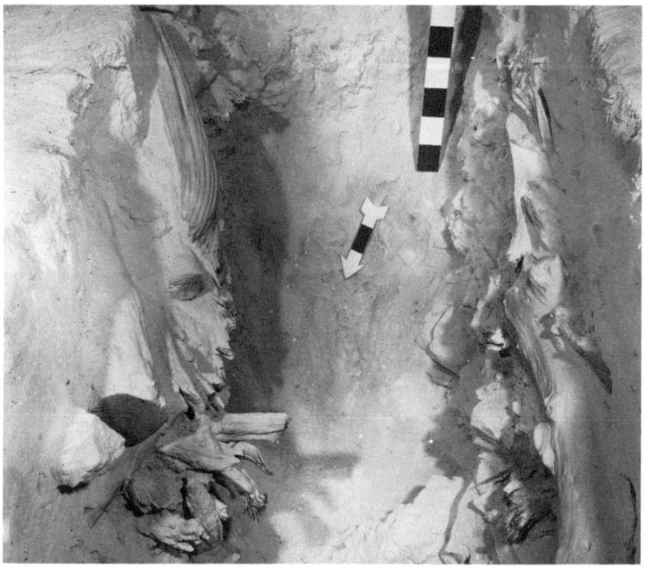

Figure 2-9. Grids O-P-Q-19 in North-South Trench showing driftwood above bedrock. North arrow is 30 cm long: vertical scale is graduated in 10 cm increments.

Figure 2-10. West wall of grids O-19 and P-19, North-South Trench, from surface of cave floor (above) to bedrock. North arrow is 30 cm long; vertical scale graduated in 10 cm increments.

Figure 2-11. East-West Trench looking west. North arrow is 30 cm long; vertical scale is 1 meter.

Figure 2-12. Grids GG and HH in East-West Trench with driftwood exposed at the 50 cm level. North arrow is 30 cm long.

from a consolidated pack rat midden 1 meter west of F.S. #3; no other provenience.

F.S. #6—A fragmentary figurine (C:5:3.14) from the surface of Antechamber 1 under a low ledge surrounded with pack rat dung; no other provenience.

F.S. #7 — A complete figurine (C:5:3.15) recovered from a small vertical crevice filled with pack rat dung 3 meters west of F.S. #5; no other provenience.

F.S. #8—An amorphous figurine fragment (C:5:3.16) recovered from a crevice between fallen rocks at a point where Antechamber 1 abuts the main room of the cave; no other provenience.

F.S. #9—Two complete figurines (C:5:3.17 and .18), one with hornlike appendages (.18), tightly wedged under a rock on the west side of the main room; surface of the cave floor; probably *in situ*.

F.S. #10 — A fragmentary figurine (C:5:3.19) under a rock at the west edge of the main room; other fallen rock around it, otherwise on the surface of the cave floor and probably *in situ*.

F.S. #11 — A fragmentary, burned figurine (C:5:3.20) in fill of a pack rat midden; no other provenience.

F.S. #12—One complete but burned figurine (C:5:3.21) in fill of the same pack rat midden as specimens 11, 13, and 20; no other provenience.

F.S. #13—One virtually complete figurine (C:5:3.22) from the surface of a pack rat midden in the upper end of Antechamber 1; no other provenience.

F.S. #14 — A fragmentary figurine (C:5:3.23) recovered from a burned pack rat midden 30 cm north-northeast of F.S. #12; no other provenience.

F.S. #15 — One complete figurine (C:5:3.24) recovered from the southeast quadrant of the main room under a large rock slab; probably *in situ*.

F.S. #16 — One complete figurine (C:5:3.25) one

Figure 2-13. South wall of grids EE and FF in East-West Trench from surface of cave floor to bedrock. Note white powdery layer. North arrow is 30 cm long; vertical scale graduated in 10 cm increments.

Figure 2-14. Vertical view of Grid I-I to the 60 cm level. North arrow is 30 cm long.

Figure 2-15. Vertical view of Grid I-I to the 1 meter level.

meter southeast of F.S. #15 in main room under a large rock slab; probably *in situ*.

F.S. #17 — A cache consisting of three virtually complete figurines (C:5:3.26, .27, and .28) on a small flat rock under a larger rock; definitely *in situ* (Figure 2-16).

F.S. #18 — One complete figurine (C:5:3.29) recovered from under a rock slab in the surface dust of the cave floor, 1 meter east of F.S. #17; definitely *in situ*.

F.S. #19 — A cache of four virtually complete figurines (C:5:3.30, .31, .32, and .33) recovered from under a rock slab 50 cm east of F.S. #18. The distal end of one figurine was protruding from surface dust; the others were under no more than 5 cm of loose dust; definitely *in situ* (Figure 2-17).

F.S. #20 — A single virtually complete figurine (C:5:3.34) recovered from under 3 cm of dust and pack rat dung in the northwest quadrant of the main room; probably *in situ*.

F.S. #22 — A single complete figurine (C:5:3.35) recovered from a pack rat midden in the northwest quadrant of the main room; no other provenience.

F.S. #23 — A single complete figurine (C:5:3.36) in association with two additional fragments (C:5:3.37 and .38) recovered from a pack rat midden 15 cm east of F.S. #22; no other provenience.

F.S. #24 — Cache of 3 figurines (C:5:3.39, .40, and .41), two complete and one fragmentary, recovered from surface of cave floor but covered with 2 cm of dust and large rocks. All were deposited upon smaller rocks; all definitely *in situ* (Figure 2-18).

F.S. #25 — Cache of three virtually complete and fragmentary figurines (C:5:3.42, .43, .44, and .45) recovered from under a large rock on the surface of the cave floor at the cliff wall; all definitely *in situ* (Figure 2-19).

F.S. #26 — Cache of two virtually complete figurines and one head fragment (C:5:3.46, .47, and .48) recovered from surface of cave floor although covered with dust. All definitely *in situ* adjacent to F.S. #25 (Figure 2-19).

F.S. #27 — Cache of two figurines (C:5:3.49 and .50), one with head missing, on top of a complete specimen, recovered from near the cliff wall in the main room 5 cm below the surface of the cave floor. *In situ*.

F.S. #28 — A fragmentary figurine (C:5:3.51) and miscellaneous fragments of another (.52) recovered from the northwest quadrant of the main room against the north wall. Imbedded in pack rat dung and small rock spalls; no other provenience.

F.S. #29 — A fragmentary figurine (C:5:3.53) recovered from a pack rat midden 1 meter west of F.S. #28; no other provenience.

F.S. #30 — A complete figurine (C:5:3.54) recovered from the northwest quadrant of the main room 2 meters south of F.S. #29 at the base of a large boulder. Specimen was 4 cm below the surface of the cave floor, probably *in situ* (Figure 2-20).

F.S. #31 — A complete figurine (C:5:3.55), pierced through the body with a split twig, recovered from the northwest quadrant of the main room at the corner of a large rock fall. Specimen was 1 to 2 cm below the surface of the cave floor; definitely *in situ*.

F.S. #32 — A complete figurine (C:5:3.56) recovered from the southwest quadrant of the main room at the base of a large rock fall. Specimen was 2 cm below the surface of the cave floor; definitely *in situ*.

F.S. #33 — A fragmentary figurine (C:5:3.57) recovered from the southwest quadrant of the main room adjacent to a large rock protruding from the cave floor. There was evidence that this area had previously been disturbed by human agency; no other provenience.

F.S. #34 — A complete figurine (C:5:3.58) recovered from the southwest quadrant and 5 cm from the west wall of the main room, 30 cm north of F.S. #35. Specimen was lying on its head and under 2 cm of cave dust. Definitely *in situ* and probably in association with F.S. #35 (Figure 2-21).

F.S. #35 — A complete figurine (C:5:3.59) placed upside down and lying on a clump of unidentified grass. Several twigs of split and unsplit willow and an unworked calcite crystal were in association. The specimen was under 2 to 3 cm of cave dust and behind a driftwood log that was in position before the figurine was deposited. Definitely *in situ* and probably in association with F.S. #34 (Figure 2-21).

F.S. #36 — Fragment of a figurine (C:5:3.60) with other fragments recovered from the northeast sector of the main room, 2 cm below the surface of the cave floor. Fragment of head indicated two wrapped hornlike appendages. Probably not *in situ*.

F.S. #37 — One complete (C:5:3.61) and one fragmentary (C:5:3.79) figurine. The distal end of the complete specimen had been inserted in the surface dirt of the cave floor; probably *in situ*.

F.S. #38 — One fragmentary figurine (C:5:3.62) recovered 20 cm south from a bedrock outcrop on the surface of the cave floor; probably not *in situ*.

F.S. #39 — One complete but fragile figurine (C:5:3.63) recovered from 2 cm beneath the surface of the cave floor between grids H-19 and I-19; probably not *in situ*.

F.S. #40 — One fragmentary figurine (C:5:3.64) recovered from 1.5 cm beneath the surface of the cave floor at the edge of Grid H-19 and 25 cm east of F.S. #39; probably not *in situ*.

F.S. #41, 42, and 43 — One complete (C:5:3.66) and three fragmentary (C:5:3.65, .67, and .68) figurines grouped together, recovered from 5 cm beneath the surface of the cave floor in Grid GG; definitely *in situ*.

F.S. #44 — One complete figurine (C:5:3.69) recovered from the disturbed surface of the northwest quadrant of the main room, 15 cm southwest of F.S. #28; probably not *in situ*.

F.S. #45 — (Identical with F.S. #49) — One complete figurine (C:5:3.73) protruding from the north wall of Grid FF, 5 cm below the surface of the cave floor; probably *in situ*.

F.S. #46 — One complete figurine (C:5:3.70) recovered from the south edge of Grid G-19, 2 cm below the surface of the cave floor between small limestone rocks; definitely *in situ*.

F.S. #47 — One complete figurine (C:5:3.71) recovered from 10 cm below the surface of the cave floor at the south edge of Grid L-19, 30 cm from the east wall of the grid; probably *in situ.*

F.S. #48 — One complete figurine (C:5:3.72) recovered from 5 cm below the surface of the cave floor at the north edge of Grid L-19, 50 cm from the east wall in grayish white calcium dust; probably *in situ.*

F.S. #49 — Identical with F.S. #45.

F.S. #50 — One fragmentary figurine (C:5:3.74) recovered from 3 cm below the surface of the cave floor in the northeast corner of Grid DD. Several pellets of what appeared to be *Ovis canadensis* dung had been inserted in the body "cavity" of the specimen; probably *in situ.*

F.S. #51 — One fragmentary figurine (C:5:3.75) recovered from 8-10 cm below the surface of the cave floor in the center of Grid M-19 in calcium dust; probably *in situ.*

F.S. #52 — One fragmentary figurine (C:5:3.76) recovered from 2 cm below the surface of the cave floor in Grid BB, 23 cm from the south wall and 5 cm from the east wall of the grid. Small limestone rocks around specimen; probably *in situ.*

F.S. #54 — One fragmentary figurine (C:5:3.77) recovered from 2 cm below the surface of the cave floor near the south wall of the main room; no other provenience.

F.S. #55 — One amorphous fragment of a wrapped twig (C:5:3.78) recovered from the surface of Antechamber 1, 10 cm south of F.S. #8.

F.S. #56 — One fragmentary figurine (C:5:3.82) recovered from 9 cm below the surface of the cave floor in Grid I-I; probably *in situ.*

F.S. #57 — One complete figurine (C:5:3.83) and 10 fragments of others (C:5:3.84 and .85) recovered from a pack rat midden; no other provenience.

F.S. #58 — One complete figurine (C:5:3.92) recovered from 10 cm below the surface of the cave floor under a 5 cm thick rock, 70 cm east of Grid I-I; undoubtedly *in situ.*

Figure 2-16. Cache of 3 figurines (F.S. #17) *in situ.* North arrow is 30 cm long.

Summary of Figurines and Figurine Caches

The data we have from those figurines and caches recovered *in situ* indicate that the makers placed them randomly on the then surface of the cave floor, usually under surficial rock fall. They were often grouped rather than placed singly. Insertion of a pellet of artiodactyl dung in the body "cavity" was rare. Piercing of the artifact with a split twig, perhaps representing a spear, was not the rule.

Subsequent to initial deposition, many figurines were moved from their original location or in other ways disturbed by rodent action.

Figure 2-17. Cache of 4 figurines (F.S. #19) *in situ.* North arrow is 30 cm long.

Figure 2-18. Cache of 3 figurines (F.S. #24) *in situ.* North arrow is 30 cm long.

Figure 2-19. Two caches of figurines (F.S. #25 left and F.S. #26 right) *in situ*. North arrow is 30 cm long.

Figure 2-20. Single figurine (F.S. #30) probably *in situ*. North arrow is 30 cm long.

Artifacts Other Than Figurines

A few miscellaneous prehistoric artifacts were recovered during the project. These were *not* in association with the figurines.

Twelve fragmentary pieces of cordage, all but one spun from yucca fiber, were found. Two fragments (C:5:3.93 and 101) came from the talus at the mouth of the cave. They were respectively, 2 ply-Z twist and 2 ply-S twist; the latter may have been of milkweed (*Asclepias* sp.) or *Apocynum* fiber. From the surface of the cave floor came two more fragments (C:5:3.88 and 91). The former was 2 ply-S twist while the latter was 2 ply-Z twist. This specimen (Figure 2-22), recovered from the

---

Figure 2-21. Figurines *in situ*, F.S. #34 on right and F.S. #35 on left. The latter specimen rests on a clump of grass in association with willow twigs behind a driftwood log. North arrow is 30 cm long.

Figure 2-22. Yucca cordage (C:5:3.91) tied with two square knots and one granny.

extreme rear of the cave, was tied with two square knots and one granny. It may have once held a group of sticks used for a torch since several burned sticks were found beneath the cordage. The remaining cordage specimens came from excavated grids:

C:5:3.87 — finely twisted, 2 ply-S twist, Grid I-I, 5-10 cm level.

C:5:3.94 — 2 ply-S twist; Grid Q19, 10 cm level.

C:5:3.95 — 2 ply-Z twist, overhand knot at one end; Grid HH, 5-10 cm level.

C:5:3.96 — 2 ply-S twist, overhand knot near mid-point; Grid HH, 5-10 cm level.

C:5:3.97 — unspun fiber square knot; Grid HH, 5-10 cm level.

C:5:3.98 — 2 ply-S twist; Grid AA, 5-10 cm level.

C:5:3.99 — 4 ply-Z twist, ends charred; Grid AA, 5-10 cm level.

C:5:3.100 — 2 ply-Z twist square knot; Grid AA, 15-20 cm level.

From the 5-10 cm level of Grid HH came a small fragmentary wooden artifact (C:5:3.102). Two edges and both sides were ground smooth and a hole had been drilled through the center. This was probably a spindle whorl.

Three olivella shell beads (C:5:3.86) came from the surface of Grid I-I. On each, the tips and bases had been ground to enlarge the openings.

A possible end scraper (C:5:3.90) came from the 10-15 cm level of Grid I-I. Large flakes had been removed unifacially along one edge. No secondary retouching was observed although use scars were present. The material was white quartzite with jasper intrusions.

Finally, a "one hand" mano (C:5:3.89) was found at the 10 cm level of Grid I-I. This was ovoid, a river cobble of fine grained reddish sandstone with two opposing use surfaces, one with evidence of striation. The specimen was complete, 11.0 cm long, 7.5 cm wide, and 3.3 cm thick.

All of these miscellaneous artifacts were probably once in association with the Pueblo II (A.D. 1050-1150) Kayenta Anasazi ruin on a terrace at the mouth of South Canyon a few hundred meters upstream from Stanton's Cave. No Anasazi *occupation* was noted in the cave and these artifacts represent occasional exploratory visits by the later Kayenta people.

From an archaeological perspective then, Stanton's Cave was the scene of ritual deposition of figurines by Archaic peoples as yet unidentified but probably related to the Pinto Basin Culture some 3000 to 4000 years ago. Later, ca. A.D. 1050-1150, Kayenta Anasazi occasionally explored the cave near their own settlement. And, in 1889, Robert Stanton's river party stored some of their gear there. There is no evidence that humans ever actually lived in the cave.

Appendices

Appendix 2-A. Description of figurines recovered under controlled scientific conditions in 1963, 1969, 1970, and 1976.

Catalog #	
C:5:3.1	Complete figurine, 12.0 cm long, 11.6 cm high, 4 body wraps, 4 neck wraps; recovered under rock fall inside left portion of main room near mouth of cave (Figure 2-A-1).
C:5:3.2	Complete figurine, 11.5 cm long, 11.0 cm high, 5 body wraps, 4 neck wraps; recovered under rock fall inside left portion of main room near mouth of cave (Figure 2-A-2).
C:5:3.3	Complete figurine, 7.7 cm long, 9.5 cm high, 3 body wraps, 5 neck wraps; recovered under rock fall inside left portion of main room near mouth of cave (Figure 2-A-3).
C:5:3.4	Complete figurine, 11.2 cm long, 11.4 cm high, 4 body wraps, 3 neck wraps; willow twig 13.5 cm long inserted through body near head and neck; recovered under rock fall inside left portion of main room near mouth of cave (Figure 2-A-4).
C:5:3.5	Complete figurine, 14.9 cm long, 12.1 cm high, 4 body wraps, 6 neck wraps; willow twig 15.0 cm long inserted through body; recovered under rock fall inside left portion of main room near mouth of cave (Figure 2-A-5).
C:5:3.6	Complete figurine, 7.5 cm long, 9.2 cm high, 4 body wraps, 7 neck wraps; split twig 26.5 cm long inserted obliquely through body; recovered from under rock fall inside left portion of main room near mouth of cave (Figure 2-A-6).
C:5:3.7	Complete figurine, 13.3 cm long, 13.1 cm high, 4 body wraps, 5 neck wraps; willow twig 8.9 cm long inserted obliquely through body; recovered from under rock fall inside left portion of main room near mouth of cave (Figure 2-A-7).
C:5:3.8	Complete figurine, 7.8 cm long, 10.0 cm high, 5 body wraps, 6 neck wraps; willow twig 16.8 cm long inserted obliquely through body; recovered from under rock fall inside left portion of main room near mouth of cave (Figure 2-A-8).
C:5:3.9	Complete figurine, 13.5 cm long, 10.5 cm high, 5 body wraps, 4 neck wraps; willow twig 21.5 cm long with bark removed inserted obliquely through body; recovered from under rock fall inside left portion of main room near mouth of cave (Figure 2-A-9).
C:5:3.10	Complete figurine, 19.0 cm long, 14.5 cm high, 5 body wraps, 4 neck wraps; willow twig 31.0 cm long inserted obliquely through body; recovered from under rock fall inside left portion of main room near mouth of cave (Figure 2-A-10).
C:5:3.11	Complete figurine, 8.5 cm long, 8.3 cm high, 6 body wraps, 5 neck wraps; artiodactyl dung pellet had been placed in body cavity during manufacture of the figurine; recovered from pack rat nest in main room of cave (Figure 2-A-11).
C:5:3.12	Fragmentary figurine, neck and head missing; body with 2 wraps was 16.6 cm long and 12.5 cm high; found wedged between rocks in Antechamber 1.
C:5:3.13	Complete figurine, 9.4 cm long, 9.0 cm high, 6 body wraps, 5 neck wraps; artiodactyl dung pellet in body cavity; recovered from pack rat nest in main room 1 m west of C:5:3.11 (Figure 2-A-11).
C:5:3.14	Fragmentary figurine (hind leg missing), neck and body wraps broken, 16.4 cm long, 19.5 cm high; recovered from under a rock ledge in Antechamber 1 (Figure 2-A-11).
C:5:3.15	Complete figurine, 5.1 cm long, 7.1 cm high, 5 body wraps, 2 neck wraps; recovered from small crevice in main room (Figure 2-A-12).
C:5:3.16	Fragmentary figurine recovered from Antechamber 1.
C:5:3.17	Complete but fragile figurine, 9.8 cm long, 8.6 cm high, 4 body wraps, 3 neck wraps; recovered from under a rock at the west side of the main room of the cave (Figure 2-A-12).
C:5:3.18	Complete figurine, 7.9 cm long, 12.0 cm high, 7 body wraps, 8 neck wraps; head wrap separate from neck wrap; two small hornlike appendages make up part of head wrap; recovered with C:5:3.17 (Figure 2-A-12).
C:5:3.19	Fragmentary figurine (head and neck missing), body is 6.3 cm long, 8 body wraps; recovered from jumbled rock at west edge of main room.
C:5:3.20	Fragile, fragmentary, and burned figurine, neck and head

missing, body is 6.7 cm long, 4 partial body wraps; recovered from pack rat nest in main room.

C:5:3.21 Complete, slightly charred figurine, 9.0 cm long, 8.6 cm high, 4 body wraps, 3 neck wraps; recovered from same pack rat nest as figurines 11, 13, and 20 (Figure 2-A-13).

C:5:3.22 Complete figurine except for portion of hind leg, 7.9 cm long, 8.7 cm high, 5 body wraps, 4 neck wraps, 2 vertical wraps around body at hind leg; recovered from pack rat nest in upper end of Antechamber 1 (Figure 2-A-13).

C:5:3.23 Fragmentary figurine consisting only of the front portion of the body and neck; recovered from pack rat nest in main room of cave.

C:5:3.24 Complete figurine, 8.5 cm long, 10.0 cm high, 4 body wraps, 4 neck wraps; recovered from the southeast quadrant of the main room of cave (Figure 2-A-13).

C:5:3.25 Complete figurine, 5.6 cm long, 8.5 cm high, 5 body wraps, 4 neck wraps; recovered from under a rock slab in the southeast quadrant of the main room of cave (Figure 2-A-13).

C:5:3.26 Fragmentary figurine (neck wraps missing), 9.9 cm long, 17.8 cm high, 7 body wraps; recovered from top of small rock under a large rock slab in the main room of cave (Figure 2-A-14).

C:5:3.27 Complete figurine, 11.5 cm long, 13.7 cm high, 5 body wraps, 6 neck wraps; recovered with C:5:3.26 (Figure 2-A-14).

C:5:3.28 Complete figurine, 10.3 cm long, 11.9 cm high, 5 body wraps, 5 neck wraps; recovered with C:5:3.27 (Figure 2-A-15).

C:5:3.29 Complete figurine, 6.1 cm long, 8.1 cm high, 5 body wraps, 6 neck wraps; recovered from under rock fall in main room of cave (Figure 2-A-15).

C:5:3.30 Complete figurine, 8.9 cm long, 12.6 cm high, 4 body wraps, 7 neck wraps; recovered from under a rock slab in the main room (Figure 2-A-15).

C:5:3.31 Complete but loosely constructed figurine, 9.9 cm long, 12.7 cm high, 6 body wraps, 1 neck wrap; recovered with C:5:3.30 (Figure 2-A-16).

C:5:3.32 Complete figurine, 11.0 cm long, 12.4 cm high, 4 body wraps, 6 neck wraps; recovered with C:5:3.30 (Figure 2-A-16).

C:5:3.33 Complete figurine, 7.6 cm long, 10.2 cm high, 6 body wraps, 5 neck wraps; recovered with C:5:3.30 (Figure 2-A-16).

C:5:3.34 Complete figurine except several body wraps broken, 7.6 cm long, 7.9 cm high, 4 body wraps still present, 6 neck wraps; recovered from northwest quadrant of main room of cave under 3 cm of dust and pack rat dung (Figure 2-A-17).

C:5:3.35 Complete figurine, 6.8 cm long, 8.3 cm high, 5 body wraps, 7 neck wraps; recovered from pack rat nest in northwest quadrant of main room of cave (Figure 2-A-17).

C:5:3.36 Complete figurine, 5.5 cm long, 8.0 cm high, 5 body wraps, 7 neck wraps; recovered from pack rat nest in the northwest quadrant of the main room of cave (Figure 2-A-17).

C:5:3.37 Fragmentary figurine; recovered with C:5:3.36.

C:5:3.38 Fragment of neck and head of figurine; recovered with C:5:3.36.

C:5:3.39 Complete figurine, 9.4 cm long, 11.4 cm high, 6 body wraps, 6 neck wraps; one hornlike appendage protruding from head; recovered from surface of main room under rock slabs with 3 other figurines (Figure 2-A-17).

C:5:3.40 Complete figurine, 9.7 cm long, 10.0 cm high, 4 body wraps, 3 neck wraps; recovered with C:5:3.39 (Figure 2-A-18).

C:5:3.41 Fragmentary figurine (hind leg missing), 15.1 cm long, 16.5 cm high, 3 body wraps, 2 neck wraps; recovered with C:5:3.39 (Figure 2-A-19).

C:5:3.42 Complete but slightly crushed figurine, 6.9 cm long, 10.5 cm high, 4 body wraps, 8 neck wraps; recovered against wall of main room with 2 other figurines and fragments (Figure 2-A-19).

C:5:3.43 Complete figurine, 12.0 cm long, 12.7 cm high, 6 body wraps, 7 neck wraps; 4 artiodactyl dung pellets inside body cavity; recovered with C:5:3.42 (Figure 2-A-20).

C:5:3.44 Fragmentary figurine, recovered with C:5:3.42.

C:5:3.45 Assorted fragments of indeterminate number of figurines recovered with C:5:3.42.

C:5:3.46 Fragment of head and neck of figurine; recovered from surface of main room of cave in association with C:5:3.47.

C:5:3.47 Complete figurine, 16.2 cm long, 15.7 cm high, 4 body wraps, no horizontal neck wraps; neck is formed by three vertical wraps extending from under chest behind foreleg to head; recovered with C:5:3.46 and .48. (Figure 2-A-20).

C:5:3.48 Complete figurine, 8.6 cm long, 11.6 cm high, 7 peeled body wraps, 8 neck wraps; unmodified twig, 27.5 cm long, inserted obliquely through two vertical wraps behind foreleg; recovered with C:5:3.47 (Figure 2-A-21).

C:5:3.49 Fragmentary figurine (head and neck missing), 5 body wraps; recovered at wall of main room of cave with C:5:3.50.

C:5:3.50 Complete figurine, 10.0 cm long, 13.0 cm high, 7 body wraps, 8 neck wraps; 2 artiodactyl dung pellets inside body cavity; recovered with C:5:3.49 (Figure 2-A-22).

C:5:3.51 Complete but loosely wrapped figurine, 10.9 cm long, 16.3 cm high, 3 body wraps and 4 neck wraps remain; recovered from north wall of northwest quadrant of main room of cave with pack rat dung, small rocks, and fragments of C:5:3.52 (Figure 2-A-22).

C:5:3.52 Fragments of a figurine in association with C:5:3.51.

C:5:3.53 Fragmentary figurine from pack rat nest in main room of cave.

C:5:3.54 Complete figurine, 8.1 cm high, 7.2 cm long, 6 body wraps, 9 neck wraps, double thickness; recovered from northwest quadrant of main room of cave (Figure 2-A-18).

C:5:3.55 Complete figurine, fragile, 11.4 cm long, 12.6 cm high, 4 body wraps, 5 neck wraps; split twig, 13.4 cm long, inserted obliquely through body; recovered from northwest quadrant of main room of cave (Figure 2-A-23).

C:5:3.56 Complete figurine, 8.2 cm long, 9.0 cm high, 4 body wraps, 5 neck wraps, 2 additional wraps around body near rear leg; recovered from southwest quadrant of main room of cave (Figure 2-A-23).

C:5:3.57 Fragmentary figurine (head and neck missing), 4 body wraps; recovered from southwest quadrant of main room of cave.

C:5:3.58 Complete figurine, 6.6 cm long, 10.6 cm high, 5 body wraps, 11 neck wraps; recovered from southwest quadrant of main room of cave (Figure 2-A-24).

C:5:3.59 Complete figurine, 9.2 cm long, 8.3 cm high, 5 body wraps, 4 neck wraps, 2 vertical wraps around rear legs; recovered near west wall of main room of cave in association with split twigs, grass, and an unworked calcite crystal (Figure 2-A-24).

C:5:3.60 Fragmentary figurine (head and neck stem only); two wrapped hornlike appendages each 3.0 cm long protruding from head; recovered from northeast quadrant of main room (Figure 2-A-24).

C:5:3.61 Complete figurine, 9.2 cm long, 12.2 cm high, 5 body wraps, 4 neck wraps; recovered from surface of main room of cave (Figure 2-A-24).

C:5:3.62 Fragmentary figurine (head missing); recovered from bedrock outcrop at rear of main room of cave.

C:5:3.63 Complete figurine except for missing neck wraps, 9.6 cm long, 11.5 cm high, 3 body wraps; recovered from a depth of 2 cm between grids H-19 and I-19 (Figure 2-A-25).

C:5:3.64 Fragmentary figurine (head and neck missing); recovered from a depth of 1.5 cm at the edge of Grid H-19.

C:5:3.65 Fragmentary figurine (head missing), 9.9 cm long, 7 body wraps; recovered with 2 other figurines from a depth of 5 cm in Grid GG.

C:5:3.66 Complete figurine, 8.6 cm long, 10.5 cm high, 8 body wraps, 8 neck wraps; recovered with C:5:3.65 (Figure 2-A-25).

C:5:3.67 Complete but fragile figurine, 11.9 cm long, 11.8 cm high; some body and neck wraps missing; recovered with C:5:3.65 (Figure 2-A-25).

C:5:3.68 Fragment of neck wrap of figurine; recovered with C:5:3.65.

C:5:3.69 Complete figurine, 7.0 cm long, 9.5 cm high, 4 body wraps,

3 neck wraps; recovered from the northwest quadrant of the main room of cave (Figure 2-A-25).

C:5:3.70 Complete figurine, 10.7 cm long, 9.7 cm high, 4 body wraps, 6 neck wraps; recovered from a depth of 2 cm at the south edge of Grid G-19 (Figure 2-A-26).

C:5:3.71 Complete figurine, 6.6 cm long, 11.4 cm high, 4 body wraps, 5 neck wraps; recovered from a depth of 10 cm at the south edge of Grid L-19 (Figure 2-A-26).

C:5:3.72 Complete figurine, 9.9 cm long, 9.5 cm high, 3 body wraps, 5 neck wraps; recovered from a depth of 5 cm at the north edge of Grid L-19 (Figure 2-A-26).

C:5:3.73 Complete, although slightly crushed, figurine, 10.4 cm long, 10.5 cm high, 5 body wraps, 5 neck wraps; recovered from a depth of 5 cm in the north wall of Grid FF (Figure 2-A-27).

C:5:3.74 Fragmentary figurine (head separated from the body), 10.4 cm long, 4 body wraps, one hornlike appendange protruding from head; recovered from a depth of 3 cm at the northeast corner of Grid DD.

C:5:3.75 Fragmentary figurine (head and neck missing), 8.9 cm long, 6 body wraps; recovered from a depth of 8-10 cm in Grid M-19.

C:5:3.76 Fragmentary figurine (foreleg, neck and head missing); recovered from a depth of 2 cm in Grid BB.

C:5:3.77 Fragmentary figurine (neck and head missing), 8.6 cm long, 5 body wraps; recovered from near south wall of main room.

C:5:3.78 Fragmentary figurine from edge of Antechamber 1 adjoining main room of cave.

C:5:3.79 Fragmentary figurine recovered in association with C:5:3.61.

C:5:3.82 Fragmentary figurine recovered from a depth of 9.0 cm in Grid I-I.

C:5:3.83 Complete figurine, 8.1 cm long, 11.1 cm high, 3 body wraps, 2 neck wraps; recovered from a depth of 5 cm in Grid I-I.

C:5:3.84 Eight fragments of an undetermined number of figurines; recovered from a depth of 5.0 cm in Grid I-I.

C:5:3.85 Two fragments of figurines; recovered from a depth of 7.5-10 cm in Grid I-I.

C:5:3.92 Complete figurine, 9.1 cm long, 10.7 cm high, 4 body wraps, 3 neck wraps; fragments of stringy bark wrapped around middle of the top portion of the back; recovered from the surface of the cave floor adjacent to Grid I-I (Figure 2-A-28).

Note: Precise proveniences of specimens are given in Chapter 2 where the above catalog numbers are associated with Field Specimen numbers and in Figure 2-2.

Figure 2-A-1. Catalog number C:5:3.1.

Figure 2-A-3. Catalog number C:5:3.3.

Figure 2-A-2. Catalog number C:5:3.2.

Figure 2-A-4. Catalog number C:5:3.4.

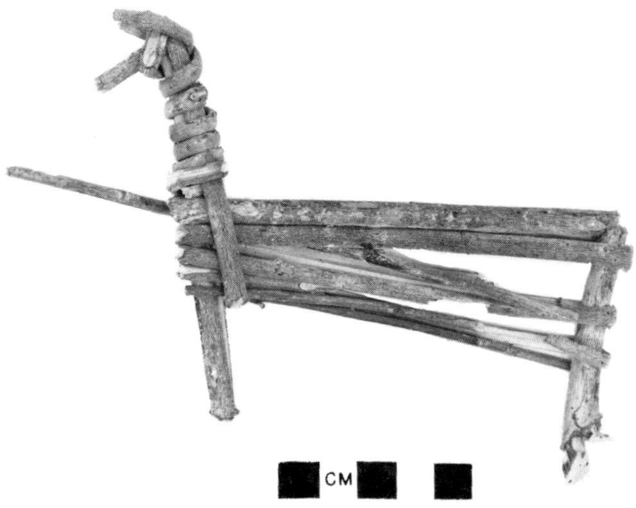

Figure 2-A-5. Catalog number C:5:3.5.

Figure 2-A-8. Catalog number C:5:3.8.

Figure 2-A-6. Catalog number C:5:3.6.

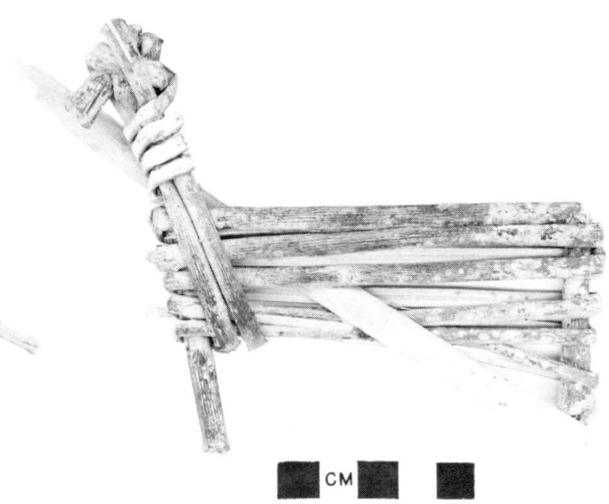

Figure 2-A-9. Catalog number C:5:3.9.

Figure 2-A-7. Catalog number C:5:3.7.

Figure 2-A-10. Catalog number C:5:3.10.

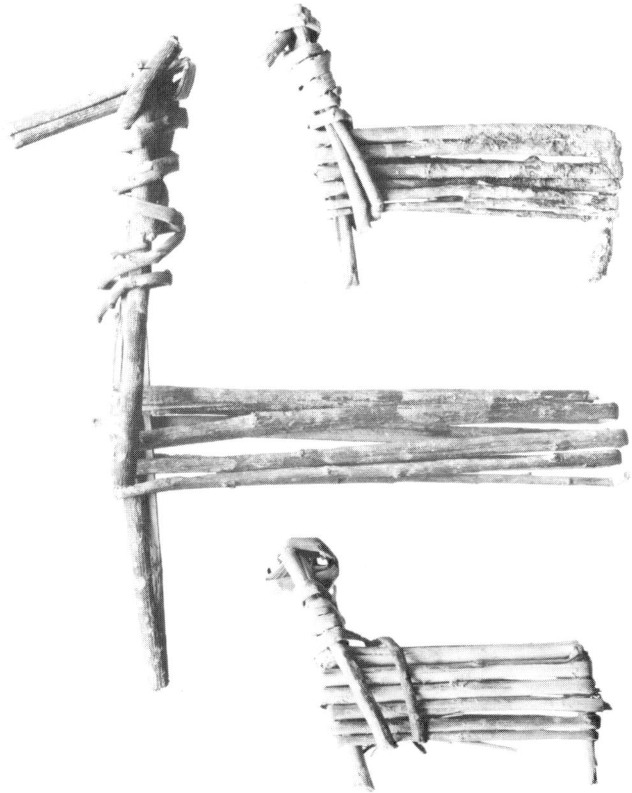

Figure 2-A-11. Catalog numbers .11, .13 and .14.

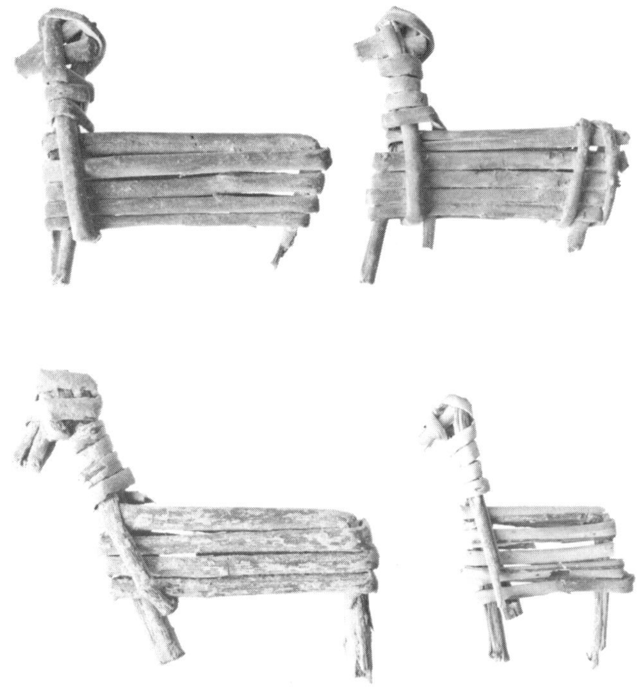

Figure 2-A-13. Catalog numbers .21, .22, .24 and .25.

Figure 2-A-12. Catalog numbers .15, .17 and .18.

Figure 2-A-14. Catalog numbers .26 and .27.

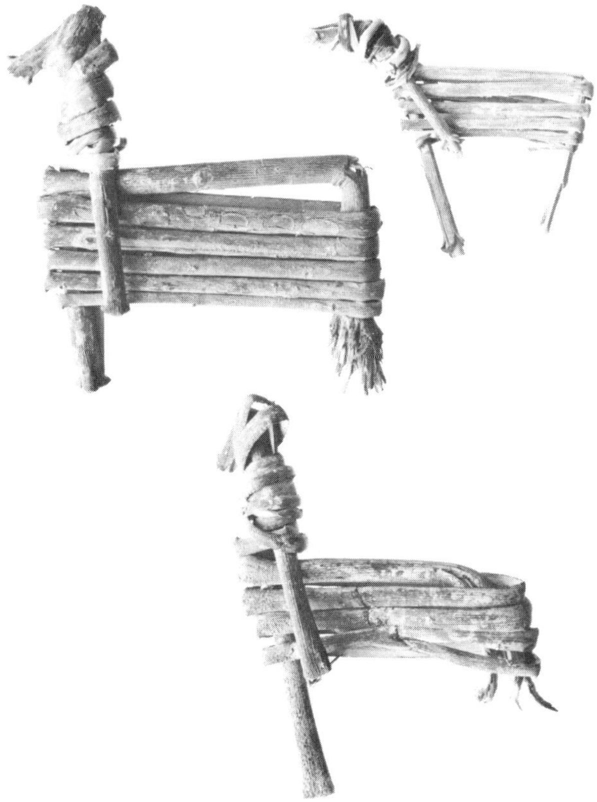

Figure 2-A-15. Catalog numbers .28, .29 and .30.

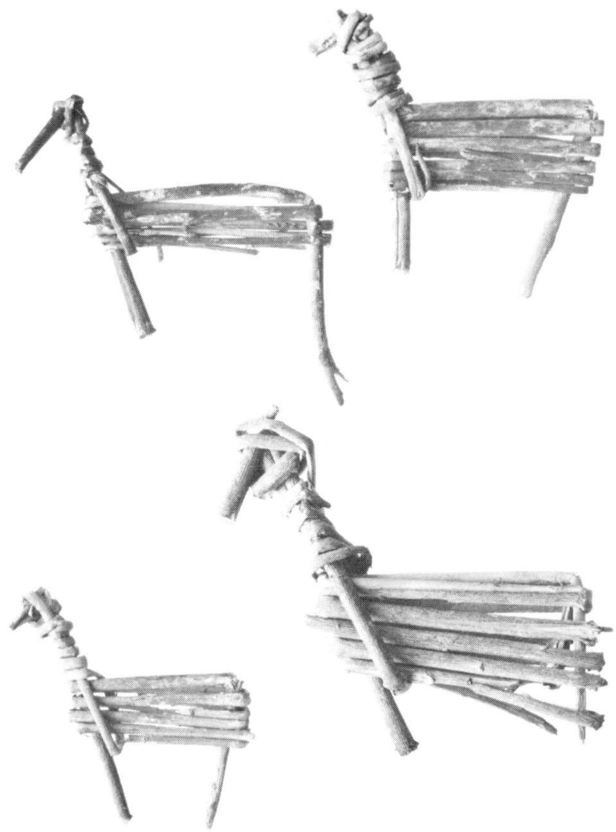

Figure 2-A-17. Catalog numbers .34, .35, .36 and .39.

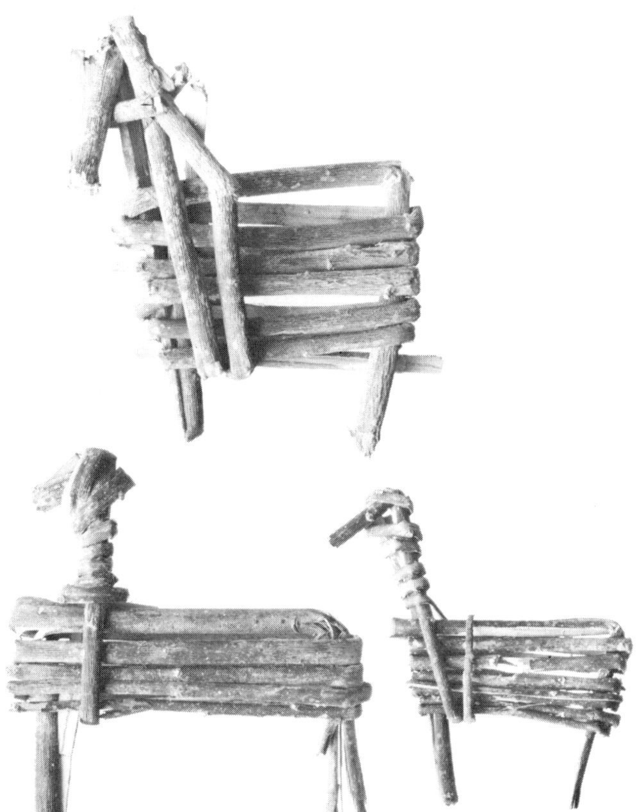

Figure 2-A-16. Catalog numbers .31, .32 and .33.

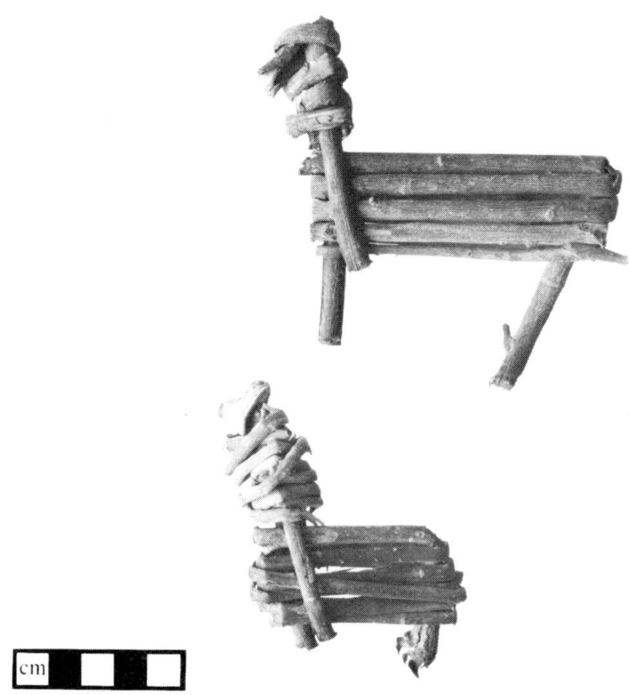

Figure 2-A-18. Catalog numbers .40 and .54.

Figure 2-A-19. Catalog numbers .41 and .42.

Figure 2-A-21. Catalog number C:5:3.48.

Figure 2-A-20. Catalog numbers .43 and .47.

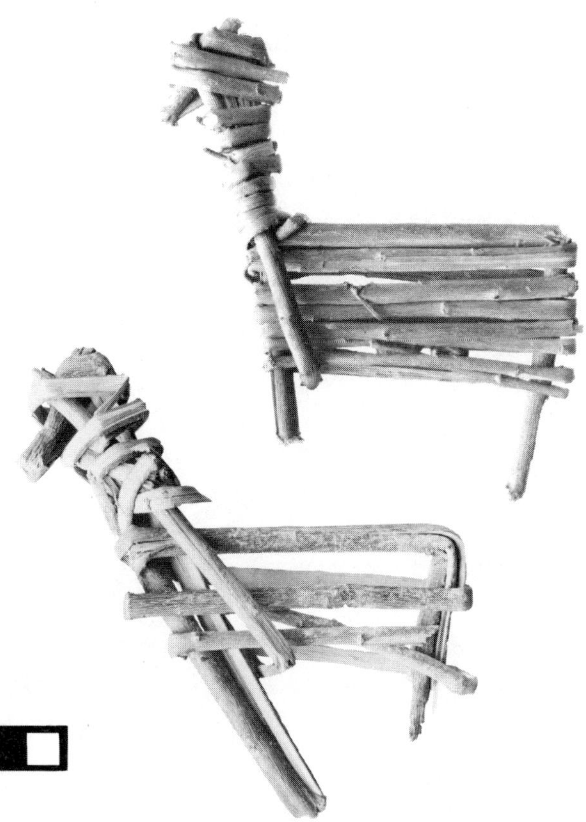

Figure 2-A-22. Catalog numbers .50 and .51.

Figure 2-A-23. Catalog numbers .55 and .56.

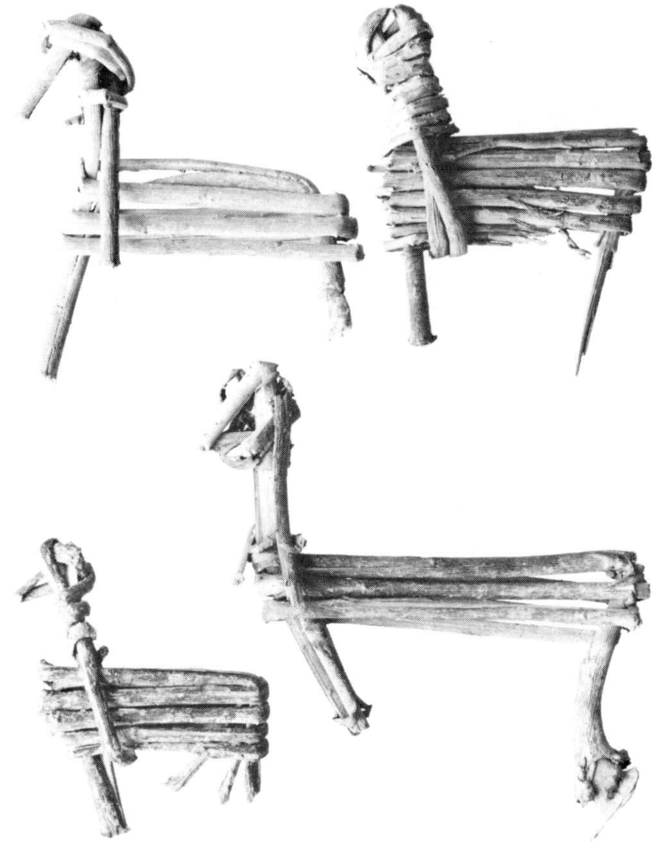

Figure 2-A-25. Catalog numbers .63, .66, .67 and .69.

Figure 2-A-24. Catalog numbers .58, .59., .60 and .61.

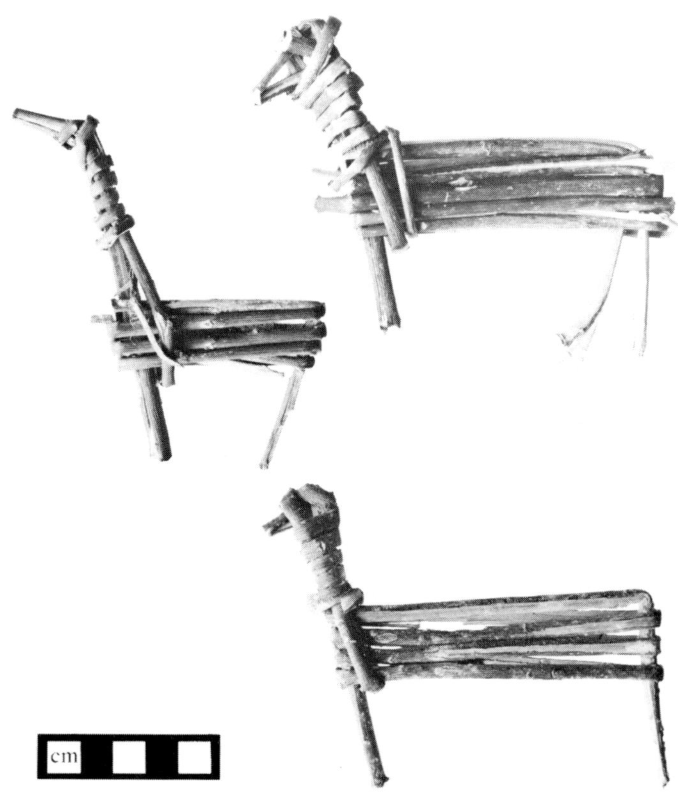

Figure 2-A-26. Catalog numbers .70, .71 and .72.

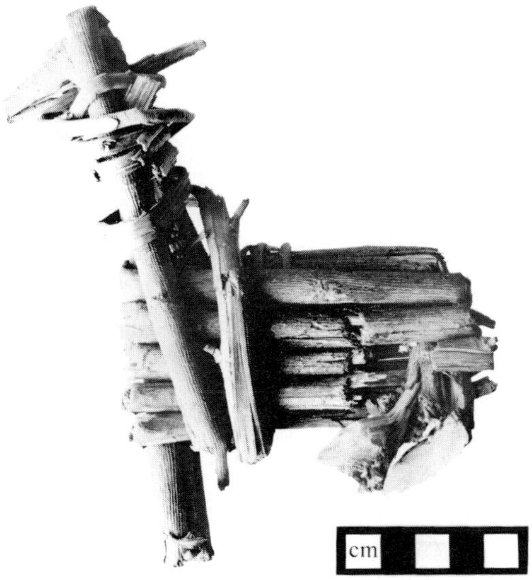

Figure 2-A-27. Catalog number C:5:3.73.

Figure 2-A-28. Catalog number C:5:3.92.

Appendix 2-B. Miscellaneous Stanton's Cave Figurines

A large group of split-twig figurines was collected illegally from Stanton's Cave in 1967. These specimens later were deposited at the Museum of Northern Arizona where they were catalogued and photographed. Subsequently, they were returned to Grand Canyon National Park where they presently are curated. Detailed descriptions of each figurine are on file at each institution.

Unfortunately, there are no definite proveniences within the cave for these figurines. Nor are the conditions of collection known. Presumably they were simply picked up wherever seen. Nonetheless, the size of the collection, consisting of 36 complete and 33 fragmentary specimens, makes discussion of value here; 44 are illustrated (Figure 2-B-1).

Most of the figurines conform to the general format of construction of these objects. Complete specimens range in size from 5.5 to 16.8 cm in height and from 5.1 to 13.7 cm in length. Average height is 10.6 cm and average length is 9.4 cm. Body wraps range from 3 to 9 with the average being 4.7.

A few specimens do show variations on the theme. One fragmentary figurine (Figure 2-B-1-F) was constructed with the addition of a grass bundle tied with willow strips extending on either side of the neck and above the head; it was attached to the neck with willow strips. Another (Figure 2-B-1-Q) had a small amount of grass and leaves pushed into the underside of the body. One figurine (Figure 2-B-1-T) had narrow strips of willow and grass knotted and looped around the hind leg as though in imitation of a snare. Another (Figure 2-B-1-CC) was recovered with a long strip of stiff bark inserted through the body cavity. On another specimen (Figure 2-B-1-FF) the neck wrapping was sufficiently long that the maker wrapped it around the body and then through one neck wrap so that it protrudes above the head. Still another figurine (Figure 2-B-1-MM) had two vertical wraps under the chest behind the forelegs, rather than the usual one, with the ends of the wraps extending above the head. These were attached to the neck by wrapping with strips of bark.

A fragmentary arrow shaft allegedly from Stanton's Cave also was included in this collection. This piece, 50.5 cm long, sinew wrapped below the nock and at the distal end of the fletching, may have come from the Anasazi site (Ariz. C:5:1) upriver. It could not have been associated with the figurines.

A final miscellaneous figurine from Stanton's Cave, also illegally collected, originally deposited at the Museum of Northern Arizona, was later transferred to the Smoki "Museum" in Prescott, Arizona. This specimen (Figure 2-B-1-SS) is 18.0 cm long, 18.7 cm high, with but two body wraps. At the head are appendages constructed by twisting a short length of vine through the head wraps. A small ball of unidentified fibre had been placed in the body cavity.

Figure 2-B-1. Miscellaneous Stanton's Cave figurines.

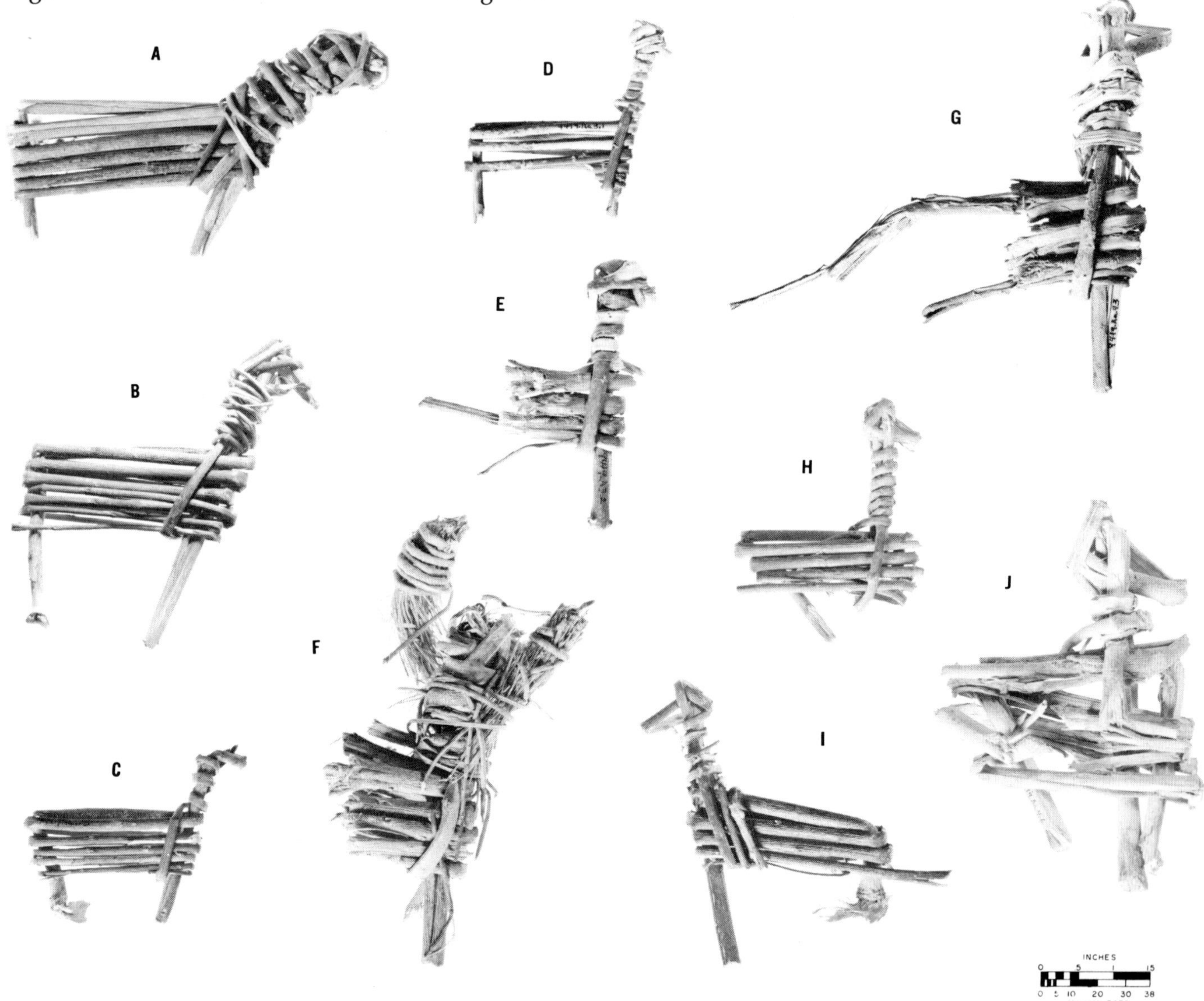

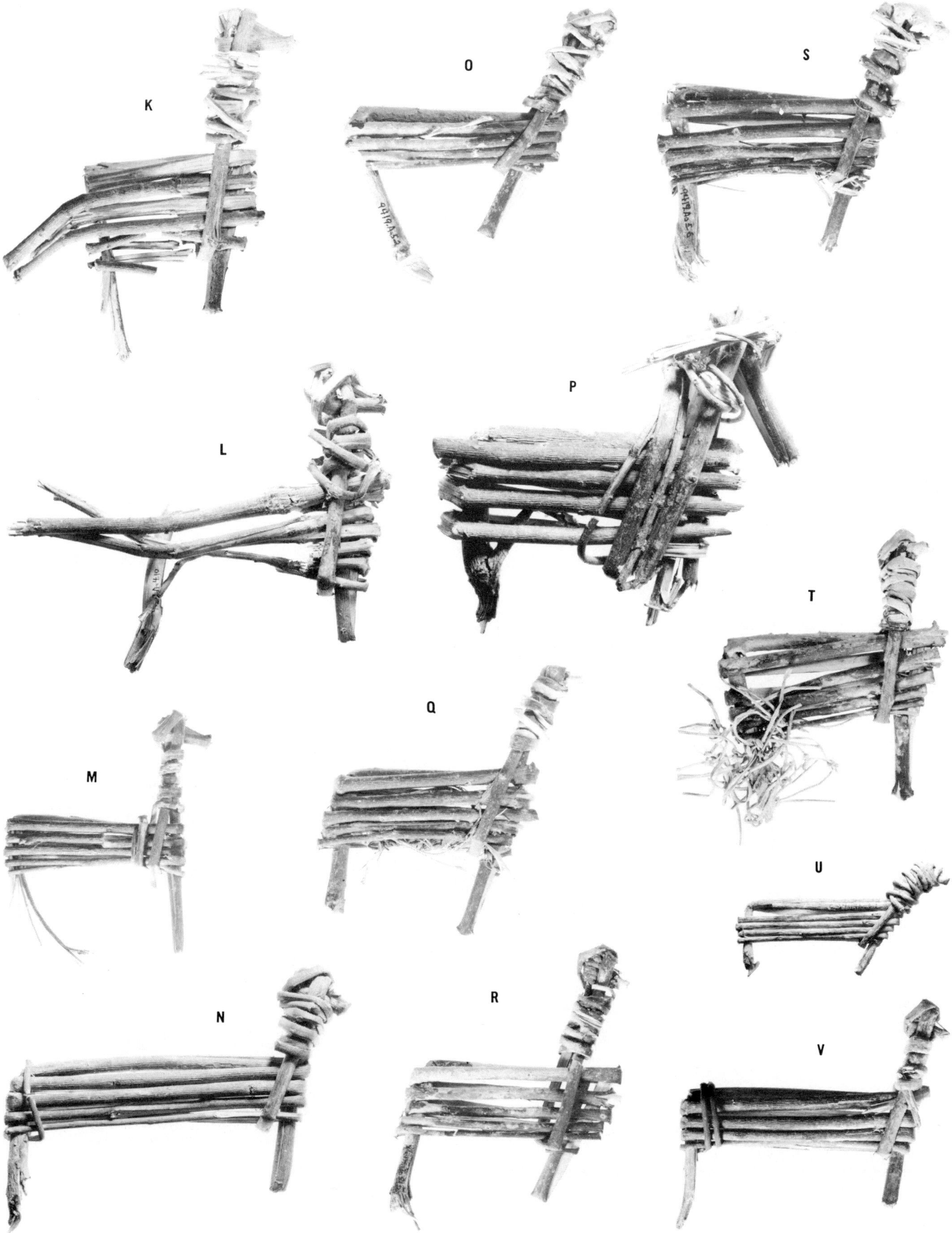
K
O
S
L
P
T
M
Q
U
N
R
V
INCHES
0 .5 1 1.5
0 5 10 20 30 38
MILLIMETERS

W
B B
G G
X
C C
H H
Y
D D
Z
E E
I I
A A
F F
J J
INCHES
0 .5 1 1.5
0 5 10 20 30 38
MILLIMETERS

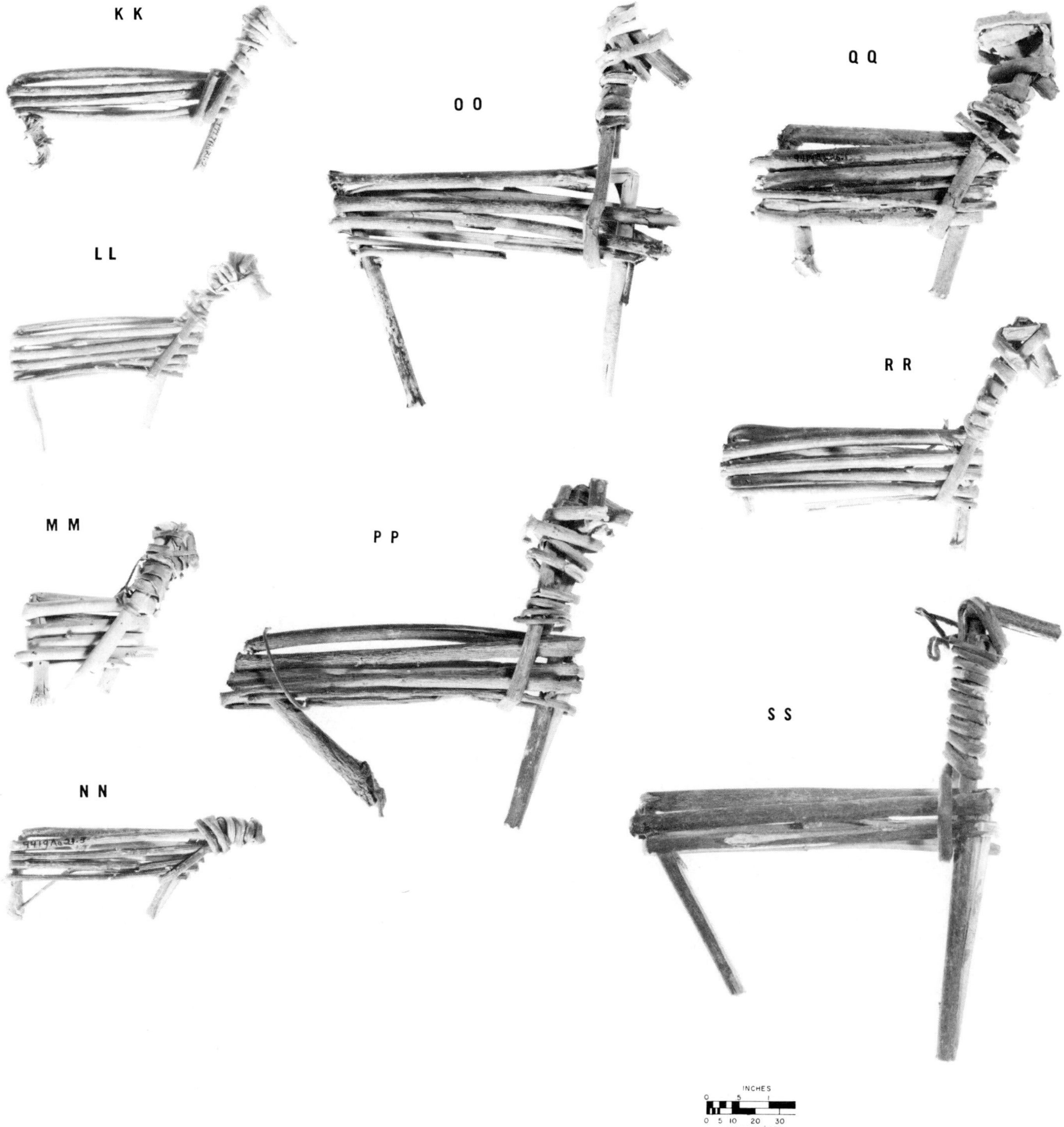
K K
O O
Q Q
L L
R R
M M
P P
S S
N N
INCHES
0 .5 1
0 5 10 20 30
MILLIMETERS

Table 2-B-1
Catalog Numbers of Miscellaneous Figurines from Stanton's Cave

MNA Catalog Number	Figure Number
NA 9419.Aa1.1	2-B-1-A
1.2	B
1.3	C
3.1	D
3.2	E
4.2	F
4.3	G
4.4	H
4.5	I
4.6	J
4.7	K
4.10	L
4.11	M
5.1	N
5.2	O
5.3	P
5.4	Q
5.5	R
5.6	S
5.7	T
6.1	U
8.1	V
10.1	W
10.2	X
11.1	Y
11.2	Z
11.3	AA
12.1	BB
13.1	CC
16.1	DD
16.2	EE
19.1	FF
19.2	GG
19.3	HH
20.1	II
21.1	JJ
21.2	KK
21.3	LL
21.4	MM
21.5	NN
22.1	OO
24.1	PP
26.1	QQ
26.2	RR
2562 Loan	SS

Footnotes

1. The figurines are described in detail in Appendix 2-A.
2. F.S. numbers not described here represent non-artifactual specimens.

References

Bohrer, Vorsila L.
1983 New Life from Ashes: The Tale of the Burnt Bush *(Rhus Trilobata)*. *Desert Plants* 5:3:122-124.

Euler, Robert C.
1966 Willow Figurines from Arizona. *Natural History* 75:3:62-67.
1983 The Pinto Basin Complex at Grand Canyon, Arizona. *The Kiva* 49:1-2:67-74.

Euler, Robert C., and Alan P. Olson
1965 Split-twig Figurines from Northern Arizona: New Radiocarbon Dates. *Science* 148:368-369.

Farmer, Malcom and R. deSaussure
1955 Split Twig Figurines. *Plateau* 27:13-23.

Jennings, Jesse D.
1980 *Cowboy Cave.* University of Utah Anthropological Papers 104. Salt Lake City: University of Utah Press.

McNutt, Charles H. and Robert C. Euler
1966 The Red Butte Lithic Sites Near Grand Canyon, Arizona. *American Antiquity* 31:3:410-419.

Schroedl, Alan R.
1977 The Grand Canyon Figurine Complex. *American Antiquity* 42:2: 254-265.

Schwartz, Douglas, A. Lange, and R. deSaussure
1958 Split-twig Figurines in the Grand Canyon. *American Antiquity* 23:264-74.

Chapter 3

Macroscopic Plant Materials From Stanton's Cave, Arizona

by

Richard H. Hevly

Department of Biology

Northern Arizona University, Flagstaff

Stanton's Cave, situated in the Mississippian Redwall Limestone approximately 44 meters above the Colorado River in Marble Canyon, has been the focus of much recent archaeological and paleontological activity (Euler 1978; Parmalee 1969). Among the abundant dessicated plant remains recovered from the 1969 and 1970 excavations were 2394 seeds, 46 small fruits, 7 flowers, 28 leaf fragments, and 23 stem or root fragments (Appendix 3-A). Miscellaneous animal material included 103 fecal pellets, 2 mollusca shells, 39 arthropod fragments or pupal cases, 6 feather fragments, and 21 bone, skin and cartilage fragments (Appendix 3-A). These materials were identified by comparison with specimens preserved in the collections of the Department of Biology at Northern Arizona University and the Museum of Northern Arizona, and consultation with Drs. Terry A. Vaughan and C. D. Johnson, Northern Arizona University. Plant names are those employed by Kearney and Peeples (1960) or McDougall (1964). The data are summarized in Table 3-1 and discussed below in relation to biogeographic and climatic changes reflected by the taxonomic composition and relative abundance of each taxa through time. Comparison where possible will be made with pollen and other paleontological and geological data.

Spatial and Temporal Relationships of Macroscopic Plant Materials and Their Transport

Macroscopic plant materials are not uniformly distributed within Stanton's Cave either horizontally or vertically (Table 3-1; Figure 3-1). Horizontal concentrations exceeding by a factor of 20 or more the concentrations in nearby areas were found in Grid EE of the East-West Trench and grids H-9 and I-19 of the North-South Trench (Table 3-1; and Appendix 3-A). Evidence of human utilization and perhaps even gathering of some of these plant materials is shown by quids of *Yucca*, charcoal, and burned plant parts found to a depth of 15-20 cm (East-West Trench only). It is not known to what extent the plant materials found in Stanton's Cave were introduced by men, but the occurrence of numerous fecal pellets including those of *Neotoma* as well as chewed seed of *Opuntia*, a favorite food of this rodent, would suggest that a major portion of the plant materials were introduced by wood rats. The horizontal concentration of seed noted above may indeed be related to the proximity of wood rat nests or selective storage by other rodents.

The highest concentrations vertically of macroscopic plant materials usually occurred approximately 5-10 cm below the surface of the cave floor and continued in many instances to a depth of 25 cm (Table 3-1; Appendix 3-A). Concentrations of seed in the vertical profiles of the trenches suggest possible relationship to periodic flooding of Stanton's Cave also indicated by concentrations of logs including these of *Juniperus*, *Pinus*, *Pseudotsuga*, and *Populus*. However, in general these log jams occur stratigraphically below the seed concentrations except in Grid J-19, in the North-South Trench and with only limited coexistence in the East-West Trench, which is elevationally lower. It may reasonably be concluded that the seeds and logs entered at different times and probably by different modes of transport. While the logs were undoubtedly introduced by water transport, the seeds were probably transported by rodents (some instances of initial gathering and utilization by man perhaps elsewhere outside of Stanton's Cave are suggested by modified plant materials, charcoal and burned plant parts). Bones, fecal pellets, teeth, hair, and arthropod fragments may also have been rodent transported but the possibility of transport by predaceous birds needs to be considered as well since owl pellets were noted by the excavators, and numerous bird bones including those of the California condor, roughlegged hawk, golden eagle, prairie falcon, and great horned owl have been reported (Parmalee 1969). Mammal bones, teeth and hair as well as arthropod fragments could also have been introduced by carnivorous mammals probably not unlike those reported by Wilson (1942) from Rampart Cave.

Seed composition (Figure 3-1) exhibits little correspondence between the East-West and North-South trenches despite superficially similar seed-rich organic deposits in both. The lithology also differs. Therefore, these seed-bearing strata are assumed to be chronologically and stratigraphically distinct provided the differences of seed content are not the result of differential sorting and concentration by rodents or other agencies. The occurrence of split-twig figurines *on the surface* of the seed-rich stratum of the North-South Trench (gray-brown humus) would suggest that the stratum is older than the upper seed-rich stratum of the East-West Trench (dark gray organic material) which *contains* split-twig figurines dating about 4000 B.P. Comparison of the seed composition of the East-West and North-South trenches with Grid I-I would also suggest that the North-South Trench record antedates that of the East-West Trench since when so arranged the trends of *Juniperus* and *Opuntia* within the trenches agrees somewhat with that observed in the test pit of Grid I-I (Figure 3-1).

The dramatic change of seed composition revealed in the sediments of Stanton's Cave is not unlike that exhibited by pollen and macroscopic plant remains analysed in Late Pleistocene deposits of the Grand Canyon and other nearby areas (Martin, Sabels, and Shutler 1961; Cole 1982; Mehringer 1967a; Phillips 1977; Spaulding and Peterson 1980; Wells and Berger 1967). Available ^{14}C dates from Stanton's Cave suggest that the change in seed composition took place since 10,760 ± 200 B.P. (Iberall 1972; Long 1983). Data from nearby caves indicates that this dramatic paleoenvironmental change had occurred by 10,500 ± 200 (Cole 1982). The pale gray-brown clayey silt of the Grid I-I and the East-West Trench, which contains an abundance of juniper seed, are, therefore, late Pleistocene in age. The gray-brown humus of the North-South Trench, which also contains an abundance of juniper seed is probably correlative. The dark gray organic material overlying the pale gray-brown clayey silt in the East-West Trench and Grid I-I has ^{14}C dates of 2450 ± 80 and 5760 ± 200 B.P., indicating the existence of a 5000 year hiatus between the two deposits, not unlike other Southwestern localities which have shown a 1000 year or more hiatus including the period generally referred to as the Altithermal (Briuer 1976; Haynes 1968; Hevly 1969; Hevly and Karlstrom 1974; Mehringer 1967b; Mehringer, Martin and Haynes 1967). The cactus-seed-rich dark gray organic deposit occupying the upper 10-20 cm of the profiles in the East-West Trench and adjoining Grid I-I was, therefore, deposited at least in part during the Little Ice Age climatic perturbations described by Haynes (1968), Hevly and Karlstrom (1974), Mehringer (1967b), Irwin-Williams and Haynes (1970) and by Baumhoff and Heizer (1965).

Biogeographic and Paleoecological Interpretation of Macroscopic Plant Data

Modern Habitat and Its Biota: Stanton's Cave is located at an elevation of 927 meters and weather data from the nearest station [Lees Ferry, Coconino County, Arizona, elevation 951 meters, 51 kilometers upstream] is indicative of a cold desert climate. Dominant plants of the area are typical northern desert shrubs such as sagebrush *(Artemisia* spp.), Mormon tea *(Ephedra* spp.), wolfberry *(Lycium andersonii)*, indigo bush *(Dalea fremontii)*, buffalo berry *(Shepherdia rotundifolia)*, matchweed *(Gutierrezia* spp.), yucca *(Yucca angustissima)*, turpentine broom *(Thamnosima montana)*, brickell-bush *(Brickellia atractyloides)* and blackbrush *(Coleogyne ramosissima)*, and numerous cacti, herbs and grasses (Woodbury 1959; Phillips 1979). In nearby more mesic or hygric areas (e.g., Vaseys Paradise, about 200 meters downstream) small trees or shrubs such as redbud *(Cercis occidentalis)*, juniper *(Juniperus osteosperma)*, cottonwood *(Populis fremontii)*, willow *(Salix* spp.), hackberry *(Celtis reticulata)*, desert almond *(Prunus fasiculata)*, arrowweed *(Pluchea sericea)*, seepwillow *(Baccharis emoryi)*, sumac *(Rhus* spp.), and Apache plume *(Fallugia paradoxa)* may be found.

Paleoecology: A total of 20 taxa were identified in the macroscopic plant remains from Stanton's Cave. Similar diversity of taxa has been shown to exist in the plant debris of late Pleistocene pack rat nests from the Mohave and Great Basin deserts and elsewhere in the Grand Canyon (Laudermilk and Munz 1934, 1938; Mead and Phillips 1981; Phillips and Van Devender 1974; Spaulding and Peterson 1980; Van Devender and King 1971; Van Devender and Mead 1976; Wells and Berger 1967). The majority of these 20 taxa are identical to those recovered by previous workers in this area of the Southwest from pack rat middens of comparable antiquity.

Most of the 20 taxa may be found within a short distance of Stanton's Cave today, but usually in more mesic habitats than that represented by the flora of the immediate vicinity of the Cave. *Pinus edulis, Juniperus osteosperma* and *Quercus gambelii* are presently found at least 610 meters above the elevation of the Cave except in riparian or narrow canyon situations where they may descend to lower elevations. Assuming that the proportion of seed types is a true reflection of local environments, it would appear that a juniper woodland surrounded Stanton's Cave during the period represented by the lower 40 cm interval of Grid I-I. (The possibility of rodent transport from driftwood on the banks of the Colorado River is also possible but still indicates changed composition of driftwood and hence altered vegetation, if not at Stanton's Cave, at least upstream.) Similar elevational depressions of plant taxa have been noted during the late Pleistocene in the Mohave and Great Basin Deserts (Martin, Sabels, and Shutler 1961; Mead 1981; Mead and Phillips 1981; Phillips and Van Devender 1974; Spaulding and Peterson 1980; Van Devender and Mead 1976).

Comparison of Macro- and Micro-Plant Data and Paleoecology of Stanton's Cave

Both macro- and micro-plant materials were recovered from Grid I-I in Stanton's Cave (Figures 3-1 and 3-2). In combination, these data provide a particularly detailed record of changing plant life surrounding this site during a period of time exceeding 35,000 years. These data also provide a unique opportunity to compare three types of information for composition and inferred paleoecology.

The major arboreal pollen types recovered were pine, spruce, birch, oak and juniper, with occasionally grains of alder, maple, willow, and walnut. The cuticle data did not include any records of trees; however the seed data recorded not only pine, juniper and oak, but also hackberry and cottonwood. The dominant nonarboreal pollen types were Graminea, Cheno-Ams, *Ephedra, Artemisia,* and other Compositae. The dominant nonarboreal seed type was *Opuntia,* which was not recorded in either the pollen or cuticle records. The pollen and cuticle records were also lacking data of *Shepherdia, Purshia* and *Allionia,* but did record the presence of various Gramineae and Compositae. Both the seed and cuticle records contained *Yucca, Agave, Amelanchier* and various Leguminosae and Cruciferae, which were missing in the pollen data. The pollen and cuticle records contained *Ephedra* which was absent in the seed data.

These overlapping records are partially complementary and reflect in their composition and proportion or frequency of recovered types their modes of accumulation. The seed record consists largely of *Opuntia* and *Juniperus,* while various types of weedy herbs and grasses comprise the bulk of the cuticle content reflecting the different mammalian vectors of transport into the cave. The pollen record consists almost exclusively of wind-transported pollen types which, at least in the uppermost sediments of the cave, occur in proportions differing only slightly from those obtained in soils outside the cave; suggesting, therefore, little contamination from disintegrated fecal matter and bias induced by opportunistic dietary preferences.

Despite differences in composition and mode of

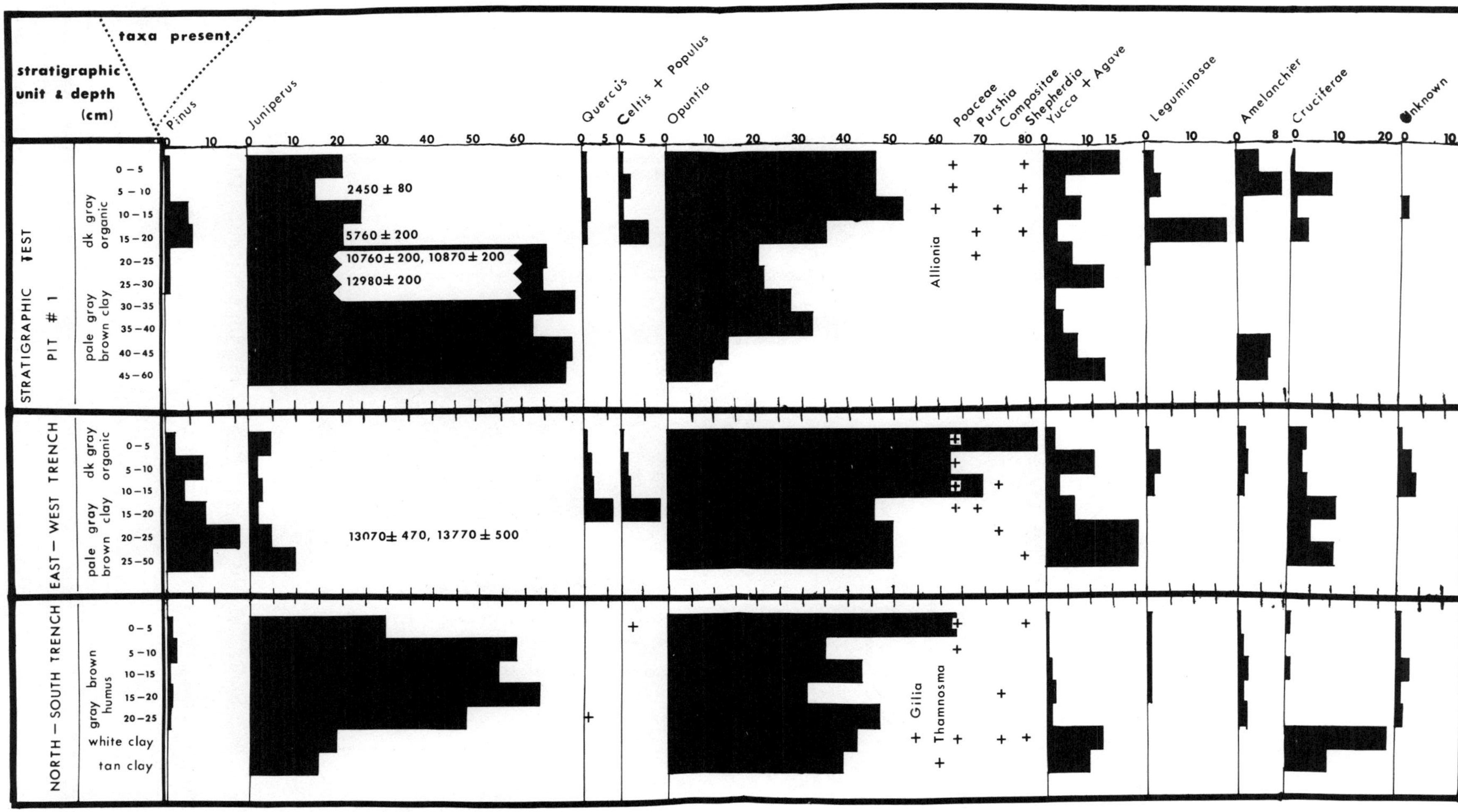

Figure 3-1. Seed composition of sediments from Stanton's Cave: Stratigraphic test pit #1, East-West Trench, and North-South Trench. (For "clay" read "clayey-silt"; for "Stratigraphic Test Pit #1" read "Grid I-I.")

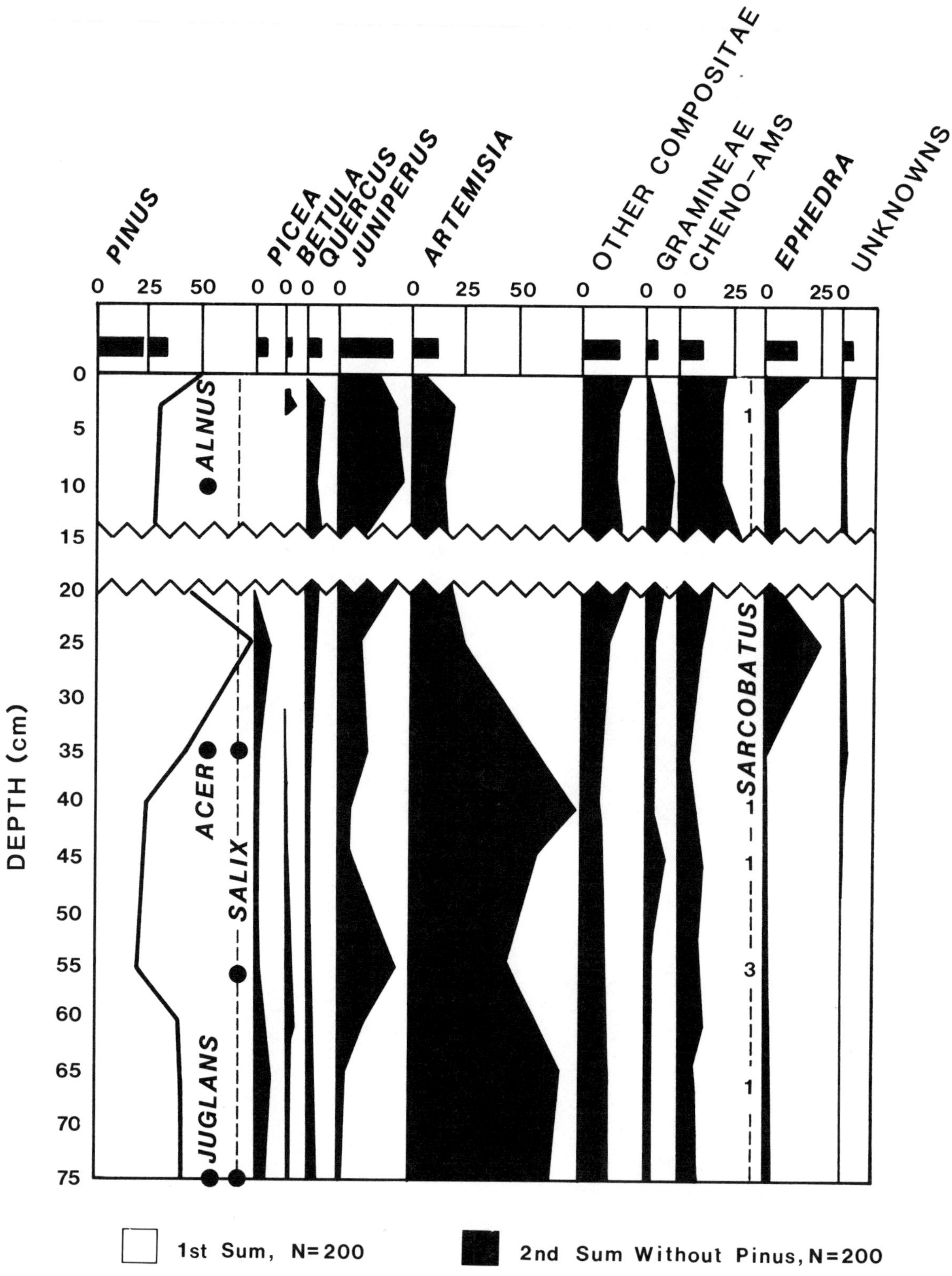

Figure 3-2. Percentages of pollen types recovered from Grid I-I compared with modern pollen sample obtained outside Stanton's Cave (uppermost spectrum). Fossil pollen data analyzed by Paul S. Martin (unpublished data used by permission). Modern pollen data from King and Sigleo (1973). Note that the surface samples of the cave deposits contain more Pinus Compositae (except Artemisia), Cheno-Ams and Ephedra pollen than do the modern soil surface samples obtained outside the cave. Juniperus, Artemisia, and Gramineae pollen percentages are higher outside the cave than inside. Similar trends have been noted elsewhere in caves, rock fissures and cisterns and have been attributed to differential transport (Hevly 1970; 1981) or differential preservation (Bradfield 1973). Since Havinga (1970) has demonstrated essentially equal preservation of pine and Juniper pollen, the former explanation is more probable. Available ^{14}C dates shown in Figures 3-1 and 3-3.

accumulation, all three records reveal changing composition of past local vegetation reflecting previous environments more mesic than the present during the late Pleistocene. Changing composition and proportion of types, particularly in the pollen and cuticle records, appear to reflect climatic perturbations of the late and post-Pleistocene which are well documented elsewhere (Appendix 3-B, Figure 3-3).

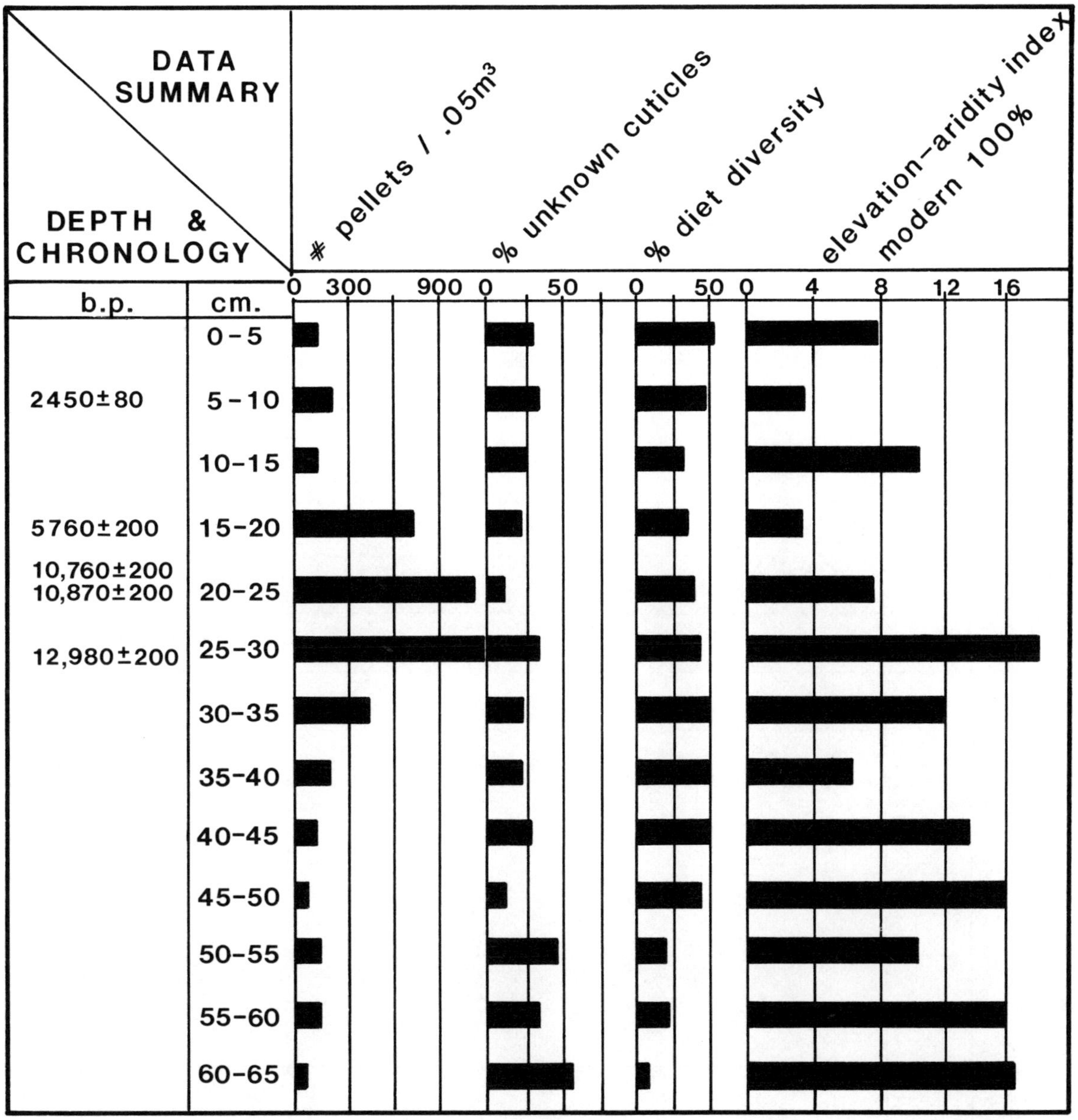

Figure 3-3. Concentration and cuticle analysis of Artyodactyl scats from Stanton's Cave (adapted from data of Iberall, 1972, summarized in Appendix 3-B). Pellets recovered below a depth of 20 cm were primarily those of Oreamnos while those recovered above 20 cm were primarily those of Ovis. The number of cuticles analyzed was the total recovered in scanning 4 horizontal and 8 vertical rows of a slide. Only the presence of a given taxon was recorded. Therefore, % Unknown Cuticle represents the proportion of unknown taxa among those examined and % Diet Diversity represents the proportion of taxa found in a particular level compared to the number of taxa found in all levels. Note the higher percentages of unknown cuticles and lower dietary diversity late in the deposit suggesting dietary change related to the dramatic Late to Post Pleistocene climatic change exhibited analysis of seed, cuticle and pollen data. Dietary trends of Ovis also exhibit a similar pattern during the Little Ice Age. The Elevation-Aridity Index was calculated for a given level by adding the products of the number of taxa found in each habitat by the previously assigned index value (see Appendix 3-B) and dividing by the total number of taxa. The above graph shows % departure from the modern value. The flux of elevation-aridity index values exhibit some parallelism with the trends shown by the pollen and seed data from Stanton's Cave and with pollen data obtained elsewhere in the Southwest (Hevly and Karlstrom 1974).

Table 3-1. Macroscopic biotic remains from Stanton's Cave.

Provenience	Grid I-I										E-W Trench							N-S Trench						
Stratigraphic Unit	Dark Gray Organic Material (1)					Pale Gray Brown Clayey Silt (2)					Dark Gray Organic Material (1)		Pale Gray Brown Clayey Silt (2)					Gray Brown Humus (3)				White Clayey Silt (4)	Tan Clayey Silt (5)	
Taxa / Depth cm	0 5	5 10	10 15	15 20	20 25	25 30	30 35	35 40	40 45	45 60	1a 0 5	1b 5 10	2a 10 15	2b 15 20	2c 20 25	2d 25 50	0 5	5 10	10 15	15 20	20 65			
Pinus	1	1	3	8	1	1					3	10	5	4	3	1	4	3	1	1	1			51
Juniperus	19	12	16	31	117	65	28	17	20	21	8	3	4	1	1	1	103	102	154	86	85	7	9	910
Quercus	1	1	1	1							1	2	3	3							1			14
*Celtis	1	1	19															1						13
Populus & Salix		1									1	1	1	1										5
*Juglans													2	3										5
*Purshia				2							1	2	1	3										9
*Shepherdia	4	2	1	2										1			1					2	4	17
Opuntia	42	37	31	52	37	22	11	9	4	3	143	83	89	21	10	5	219	60	133	43	84	121	13	1,172
Yucca & Agave	15	4	5	4	11	12	1	1	2	4	4	15	4	3	4	2	2		3	1	2	5	3	107
Legume a	2	1			1							1		1			1	1		1				9
Legume b		2		25	1						1	3	2						1			1		36
Grass Frag.	3	2									1	1	1				1	2				1		12
Comp. Frag. (cf. *Brickellia* and *Cirsium)*			1										1	1						1		1		5
Cruciferae	1	7	1	7	3				2	2	8	4	7	5	1	1	5	1	1			11	3	70
Amelanchier		8	1	2	5						3	3	2				2	1	5	1	2		1	36
Unknown 1			1								1	2	4											8
Unknown 2			1									2	1				1	1	7	1	2			16
Gilia																						1		1
Thamnosma																							1	1
Allionia				1																				1
Subtotal	89	79	63	144	176	100	40	27	28	30	175	132	127	47	19	10	339	172	305	135	177	50	34	2,498
F.P. Mammal			1	1	6		1				10	8	4	3			7	14	31	3		3	1	93
F.P. Bird	1			1																			6	8
F.P. Reptile																						2		2
Arthropod Frag.		1	2	4	3						2	2	4	2				1	2			11	5	39
Mollusc Frag.											1											1		2
Unknown Plant																								
Unknown Root																								
Unknown Stem												4	3				1	2	1	1	3		8	23
Unknown Animal																								
Unknown Feather							1												2	1		1		6
Unknown Bones				6	3	5											1		2	1	1	1	1	21

* Strata are of uneven thickness, therefore tabulation by relative postion rather than depth (e.g. upper half or ¼ *vs.* lower half or ¼).

References

Baumhoff, M.A., and R.F. Heizer
1965 "Postglacial Climate and Archaeology in the Desert West." *In* H.E. Wright, Jr., and D.G. Frey (eds.), *The Quaternary of the United States.* Princeton: Princeton University Press.

Briuer, F.L.
1976 *Plant and Animal Remains from Caves and Rock-shelters of Chevelon Canyon, Arizona.* Ph.D. Dissertation. Los Angeles: University of California.

Bradfield, M.
1973 Rock Cut Cisterns and Pollen Rain in the Vicinity of Old Oraibi. *Plateau* 46:68-71.

Cole, K.
1982 Late Quaternary Zonation of Vegetation in the Eastern Grand Canyon. *Science* 217:1142-1145.

Euler, Robert C.
1978 Archeological and Paleobiological Studies at Stanton's Cave, Grand Canyon National Park, Arizona — A Report of Progress. *National Geographic Society Research Reports, 1969 Projects* 141-162.

Havinga, H.
1971 "An Experimental Investigation into the Decay of Pollen and Spores in Various Soil Types." *In* Brooks et al. (eds.), *Sporopollenin.* New York: Academic Press.

Haynes, C.V., Jr.
1968 Geochronology and Quaternary Alluvium. In R.B. Morrison and H.E. Wright, Jr. (eds.), *Means of Correlation of Quaternary Succession.* Salt Lake City: University of Utah Press.

Hevly, R.H.
1969 "Sand Dune Cave Pollen Studies." *In* A.J. Lindsay, Jr., et al., *Survey and Excavations North and East of Navajo Mountain Utah, 1959-1962.* Museum of Northern Arizona Bulletin #45 (Glen Canyon Series #8, Flagstaff. Pp. 393-396.
1970 Botanical Studies of Sealed Jar Cached near Grand Falls, Arizona. *Plateau* 42:150-156.
1981 Pollen Production, Transports and Preservation: Potentials and Limitations in Archaeological Palynology. *Journal of Ethnobiology* 1:39-54.

Hevly, R.H., and T.N.V. Karlstrom
1974 "Southwest Paleoclimate and Continental Correlations." *In* T.N.V. Karlstrom et al. (eds.), *Geology of Arizona.* Geological Society of America (Rocky Mountain Section) Field Guide.

Iberall, E.R.
1972 *Paleoecological Studies from Fecal Pellets: Stanton's Cave, Grand Canyon, Arizona.* MA thesis. Tucson: University of Arizona.

Irwin-Williams, C., and C.V. Haynes, Jr.
1970 Climatic Change and Early Population Dynamics in the Southwestern United States. *Quaternary Research* 1:59-71.

Kearney, T.H., and R. H. Peebles
1960 *Arizona Flora.* Berkeley: University of California Press.

King, J.E., and W.R. Sigleo
1973 Modern Pollen in the Grand Canyon. *Geoscience and Man* 7:73-81.

Laudermilk, J.D., and P.A. Munz
1934 Plants in the Dung of Nothotherium from Gypsum Cave, Nevada. *Carnegie Institute of Washington Publications* 453:29-37.
1938 Plants in the Dung of Nothotherium from Rampart and Muav Caves, Arizona. *Carnegie Institute of Washington Publications* 487:271-281.

McDougall, W.B.
1964 *Grand Canyon Wildflowers.* Flagstaff: Museum of Northern Arizona.

Martin, P.S., B. Sabels and D. Shutler
1961 Rampart Cave Coprolite and Ecology of the Shasta Ground Sloth. *American Journal of Sciences* 259:102-127.

Mead, J.I.
1981 The Last 30,000 Years of Faunal History Within the Grand Canyon, Arizona. *Quaternary Research* 15:311-326.

Mead, J.I., and A.M. Phillips
1981 The Late Pleistocene and Holocene Fauna and Flora of Vulture Cave, Grand Canyon, Arizona. *Southwestern Naturalist* 26:257-288.

Mehringer, P.J., Jr.
1967a "Pollen Analysis of the Tule Spring Area, Nevada." *In* H.M. Wormington and D. Ellis (eds.), *Pleistocene Studies in Southern Nevada. Anthropological Papers* #13. Carson City: Nevada State Museum. Pp. 129-200.
1967b Pollen Analysis and the Alluvial Chronology. *The Kiva* 32:96-101.

Mehringer, P.J., Jr., P.S. Martin and C.V. Haynes, Jr.
1967 Murray Springs, a Mid-Post Pleistocene Pollen Records from Southern Arizona. *American Journal of Science* 265-786-797.

Parmalee, P.W.
1969 California Condor and Other Birds from Stanton's Cave, Arizona. *Journal of the Arizona Academy of Science* 5:204-206.

Phillips, A.M.
1977 *Packrats, Plants and the Pleistocene in the Lower Grand Canyon.* Ph.D. Dissertation, University of Arizona, Tucson.
1979 *Grand Canyon Wildflowers.* Grand Canyon: Grand Canyon Natural History Association.

Phillips, A.M., and T.R. Van Devender
1974 Pleistocene Pack Rat Middens from the Lower Grand Canyon of Arizona. *Journal of the Arizona-Nevada Academy of Science* 9:117-119.

Spaulding, W.G., and K.L. Petersen
1980 "Late Pleistocene and Early Holocene Paleocology of Cowboy Cave." Appendix II *In* J.D. Jennings (ed.), *Cowboy Cave*. Salt Lake City: University of Utah Anthropological Papers Nbr. 104.

Van Devender, T.R., and J.E. King
1971 Late Pleistocene Vegetational Records in Western Arizona. *Journal of the Arizona Academy of Science* 6:240-244.

Van Devender, T.R., and J.I. Mead
1976 Late Pleistocene and Modern Plant Communities of Shinumo Creek and Peach Springs Wash, Lower Grand Canyon, Arizona. *Journal of the Arizona-Nevada Academy of Science* 11:16-22.

Wells, P., and R. Berger
1967 Late Pleistocene History of Coniferous Woodland in the Mohave Desert. *Science* 155:1640-1647.

Wilson, R.W.
1942 Preliminary Study of the Fauna of Rampart Cave, Arizona. *Carnegie Institute of Washington Publications* 530:169-185.

Woodbury, Angus M. (ed.)
1959 *Ecological Studies of the Flora and Fauna in Glen Canyon*. University of Utah Anthropological Papers 40:1-226. Salt Lake City: University of Utah Press.

Appendix 3-A. List of macroscopic plant materials from Stanton's Cave, Arizona (1969-1970 excavations). Strata designations are described in Table 1 and shown in Figures 1 and 2. Materials which are charred or otherwise modified by man are circled. Leaves are indicated by single *, fruits by two asterisks, flower by three asterisks, and stem and root fragments by four asterisks. No symbol indicates seed. Beetle fragments are of Tenebrionidae.

Horizontal & vertical provenience (cm) & strata designation \ Species	Pinus	Juniperus	Yucca	Quercus**	Celtis	Cruciferae	Brickellia	Unknown 1	Unknown 2	Amelanchier	Purshia	Shepherdia	Leguminosae	Legumminosae	Populix & Salix	Opuntia	
Cave passageway Rats Nest to North			1* 1**	1													1 charred stem
Rats nest near FS3			1**	1													
Talus outside fence			(1*) circled														
AA level (1a)						1										1	1 rabbit fecal pellet
AA 0-5 (1a)						1		1						1		6	1 pup. case (Dipteran) 1 f. pel. fragment
AA 5-10 (2a)						4	1	1	1	1				1		5	2 pup. case (Dipteran) 1 nut. frag. 1 f.p.
AA 10-15 (2b)	1					2					2					6	
AA 15-20 (2c)			2													8	
AA 20-25 (2d)						1										2	
BB level (1a)						3										1	
BB 5-10 (1b)						1										10	
BB 10-15 (2a)			1													1 3	1 bark plate (Pinus?) 1 b. ely. 2 burnt stem f.
BB 15-20 (2b)			1**									1				1*** 8	(1 beetle ely., 2 nut. frag., 3 rab. f. pel.)
BB 20-25 (2c)	2		2			1										2	
CC 0-5 (1a)																5	2 rabbit fecal pellets
CC 5-10 (1b)	1	1						1	1	1				1		5	
CC 15-20 (2a)								1						1		5	
CC 20-25 (2b)	1					2	1				1		1				1 pup. case (Dipteran) 1 nut fragment
CC 25-50 (2c)	1	1															
DD 0-5 (1a)																8	1 shell disk pierced
DD 5-10 (1b)			5 5*	1		2			1							14	3 pieces of stem, 2 rab. f. pel., 1 f. frag.
DD 10-15 (2a)	3	1	1**	1												7	(1 rab. f. pel., 1 nut fragment)
DD 15-40 (2d)		1														1***	(1 rock)
EE 0-5 (1a)						1				1						12	2 rabbit fecal pellets
EE 5-10 (1b)	1		1*												1	12	1 grass root, 1 feather, 1 rab. f. pellet
EE 10-15 (2a)	1					1		2		1	1					63	1 rabbit fecal pellet, 1 Populus
EE 15-20 (2b)	2	1	2*	3		1									1	6	
EE 20-45 (2d)	1		2*														
FF 0-5 (1a)	1	2								1						14	
FF 5-10 (1a)			2							1						13	
FF 10-15 (1b)	3		1* (1*) circled							1				1		9	
FF 15-20 (1b)	2		1 1**			1				1	2					9	1 charcoal
FF 20-25 (2a)	1	2	1	3		1										3	1 beetle elytra, 1 rab. f. pellet, 1 rock
FF 25-50 (2d)																2***	
GG 0-5 (1a)	1	1	1 (1*) circled									1				30	1 grass
GG 5-10 (1a)		1		1											1	13	1 fecal pellet (rabbit)
GG 10-15 (1b)	1	1		1									1			13	2 rabbit fecal pellets
GG 15-20 (1b)	1													1			2 beetle elytra
GG 20-25 (2a)						1											1 grass

Horizontal & vertical provenience (cm) & strata designation \ Species	Pinus	Juniperus	Yucca	Quercus**	Celtis	Cruciferae	Brickellia	Unknown 1	Unknown 2	Amelanchier	Purshia	Shepherdia	Leguminosae	Legumniosae	Populix & Salix	Opuntia	
HH 0-5 (1a)		2	1*			1										17	1 fecal pellet (large sheep or deer)
HH 5-10 (1a)	1	2				1										23	2 rab. fecal pellets, 1 pupal case (Dipteran)
HH 10-15 (1b)	1							1								6	1 rabbit fecal pellet
HH 15-20 (1b)		1														5	1 rabbit fecal pellet
HH 20-25 (2a)			1													1** 1	
G-19 0-5 (3)		12				1										40	
G-19 5-10 (3)		1														10	
G-19 10-15 (3)		8							1	1						4** 31	1 pupal case, (Dipteran), 1 rock, 1 scorpion pincer
G-19 15-40 (3)		9	1* 1**													3** 2	
G-19 40-65 (3)		8														1	
H-19 0-5 (3)	4	28							1	1			1			98	1 rabbit fecal pellet
H-19 5-10 (3)		34			1				1	1			1			1** 22	3 rabbit fecal pellets, 1 grass fragment
H-19 10-15 (3)		63	2*						3	2						1** 40	1 root, 1 fish verti, 29 f. rat pel.(including mouse), 2 feathers
H-19 15-20 (3)	1	13											1			2**	1 Circium involucre, 1 wood frag., 1 bone, 1 rab. f. pel
H-19 20-25 (3)		13	1 **						1	1						20	
H-19 25-50 (3)	1	7														2	
I-19 0-5 (3)		2														14	
I-19 5-10 (3)	2	35								1						26	1 beetle elytra, 2 rocks, 10 fecal rat pellets
I-19 10-15 (3)	1	83	1 1*			1			3	2				1		2**** 53	1 bone, 1 rock, 2 rat fecal pellets
I-19 15-20 (3)		64							1	1						36	2 rat fecal pellets, 1 feather fragment
I-19 20-25 (3)		53	1**						1	1						57	1 bark plate (pinus p.), 2 stem/root frag., 1 fish vert.
I-19 25-50 (3)		4		1												4	
J-19 0-5 (3)		40	1* 1**			3				1		1				40	1 grass stem frag., 1 stem frag., 6 rat fecal pellets
J-19 5-30 (3)	1	32														1**	1 rabbit fecal pel., caliche nodule, 1 grass rhizome, 2 stem
K-19 0-5 (3)		21				1										25	
K-19 5-30 (3)		6														2	
K-19 0-5 (3)																2**	hide with hair-deer
L-19 0-5 (4)						1										1	1 beetle ely., 1 cartilage frag., 2 rat f. pel., 1 gilia**
L-19 10-15 (5)		2														1	2 fecal bird pellets, 2 beetle elytra
M-19 10-15 (4)						2						1				3	2 p. cases, 1 insect head, 3 b. ely., 2 liz. f. pel, 1 gastropod
M-19 15-40 (5)																	3 burnt pieces wood, 1 pith, 1 stem Thamnosma
O-19 0-5 (5)												2					1 stem fragment, 1 charcoal
O-19 5-10 (5)												1					2 bird feces, 5 rock, 1 charcoal, 1 bone
O-19 10-15 (5)		1	2*			3						1				7	3 beetle ely., 2 bird feces, 1 rat fecal pellet
P-19 5-10 (4)		7				4								1		11	3 b. ely., 1 caliche nod., 10 rocks, 1 p.c. (Dip). 1 f.p
P-19 10-35 (5)			(1*)							1						3****	1 wood (Jun.) fragment
Q-19 5-10 (4)			3				1										2 rocks, 1 feather fragment
Q-19 5-10 (4)			2			5						1				6	1 grass spiklet
I-I 0-5 (1)	1	19	11	1	1	1						4	2			(1) 41	5 caliche nod., 1 grass rhizome, 4 agave. seed, 1 rock, 1 bird feces, 1 grass spiklet, 1 grass fruit
I-I 5-10(1)	1	12	3	1		7				8		2	1	2		(1) 36	1 cottonwood fluff, 1b. thorax, 1 grass spiklet, 1 fluff grass (Tridens sp.)
I-I 10-15 (1)	3	(1) 15	5	1	1 (1)	1		1	1	1		1				31	1 dipteran p. case, 1 b. ely., 2 rocks, 1 rat f. pel., composite achene winged

Horizontal & vertical provenience (cm) & strata designation \ Species	Pinus	Juniperus	Yucca	Quercus**	Celtis	Cruciferae	Brickellia	Unknown 1	Unknown 2	Amelanchier	Purshia	Shepherdia	Leguminosae	Legumniosae	Populix & Salix	Opuntia	
I-I 15-20 (1)	8	30	4	1	9	7				2	2	2		25		50	1 Juniper berry (c.f.j. deppeana), 1 bird feces, 2 cactus spines, 4 b.e., 4 bone, 1 rat f.p., 1 packrat tooth, small immature rodent, 1 Alliona
I-I 20-25 (2)		2															3 bones (small immature rod.), 2 dip. p. cases, 6 f. pal. (rat), 1 ely.
I-I 20-25 (2)	1	115	11			3				5	1		1	1		37	2 rabbit bones
I-I 25-30 East (2)		35	11													1** 6	3 bones (small juvenile rodent)
I-I 25-30 West (2)	1	30	2													15	1 fecal pellet (rat), 1 feather fragment
I-I 30-35 (2)		28														11	
I-I 35-40 (2)		17	1													9	
I-I 40-45 (2)		20	2			2										4	
I-I 45-50 (2)		10														1	
I-I 50-55 (2)		8	3			2										1	
I-I 55-60 (2)		3	1													1	
Totals	51	910	107	14	13	70	5	8	16	36	9	17	9	36		1172	

Appendix 3-B. Cuticle identifications from Artiodactyl scats recovered from Grid GG of Stanton's Cave, Arizona (Iberall 1972), arranged according to stratigraphic occurrence and ecological distribution (1 & 2: 0-3500 feet; 3 & 4: 3500-6500 feet; 5 & 6: 6500-9500 feet). Three to six pellets per level were analyzed.

Distribution: Hygric-Mesic (2, 4, 6)

Provenience Depth (cm)	-Adiantum capillus-venerus	-Carex curatorum	Clematis ligusticifolium	-Eleocharis sp.	-Equisetum sp.	*Hedoma diffusum	Mimulus cardinalis	-Panicum huachucae	Phragmites communis	Prunus fasiculata	*Tragia stylaris	+ Urtica serra
0-5		x			x		x	x	x	x		
5-10		x						x	x	x		
10-15								x		x		
15-20	x							x		x		
20-25							x	x		x		
25-30					x	x	x	x	x	x		
0-35	x					x	x	x	x	x		
35-40			x					x	x	x		
40-45			x				x	x	x	x	x	x
45-50				x			x	x	x	x		
50-55												
55-60							x	x				
60-65						x	x					

Distribution: Xeric (1)

Provenience Depth (cm)	Aplopappus spinulosus	Aristida glauca	Aristida longiseta	Aster abatus	Astragalus sp.	Atriplex jonesii	Coldenia sp.	Cryptantha barbigera	Datura metaloides	Eriogonum deflexum	Eriogonum inflatum	Galium sp.	Mentzelia sp.	Oryzopsis hymenoides	Tridens pulchellus
0-5		x	x	x		x	x	x	x		x	x			x
5-10		x	x	x	x		x	x		x		x			x
10-15	x	x			x									x	
15-20		x	x	x			x			x					x
20-25		x	x		x			x				x	x		
25-30		x	x			x		x			x	x			x
0-35		x	x		x						x	x			x
35-40		x			x		x	x		x		x	x		x
40-45		x	x	x			x	x				x			x
45-50		x	x			x	x	x					x		
50-55							x	x							
55-60		x													x
60-65		x													

Distribution: Xeric (3)

Provenience Depth (cm)	Agave utahensis	+ Amelanchier utahensis	Artemisia tridentata	Atriplex sp.	+ Berberis repens	Bouteloua gracilis	Bouteloua sp.	Chenopodium sp.	Chrysopsis hispidum	Enneapogon desvauxii	Ephedra torreyana	Fallugia paradoxa	Laphamia congesta	Lesquerella arizonica	Lycurus phleoides	Schedonardus paniculatus	Sporobolus spp. (texanus)	Sphaeralcea sp.	Stipa arida	Yucca angustissima	Yucca baccata
0-5	x	x	x				x			x					x	x	x	x	x	x	x
5-10	x		x	x			x			x	x	x			x	x	x		x		
10-15	x			x		x		x		x	x		x		x	x	x	x			x
15-20					x		x			x					x		x	x			
20-25	x		x	x					x	x	x				x		x	x		x	
25-30	x		x				x			x					x		x	x	x		
0-35	x		x	x			x			x		x			x	x	x	x	x	x	
35-40	x		x	x			x			x					x	x	x	x	x	x	x
40-45	x		x										x		x	x	x	x	x		
45-50		x				x				x			x	x	x	x	x	x	x	x	
50-55	x									x	x		x		x		x			x	
55-60	x		x								x		x		x		x				x
60-65																					x

Distribution: Xeric (5)

Provenience Depth (cm)	Agropyron sp.	Arenaria fendleri	Solidago sp.
0-5	x	x	
5-10	x		
10-15	x		
15-20	x	x	
20-25	x		
25-30	x	x	
0-35	x		
35-40	x	x	
40-45	x		x
45-50	x		
50-55	x		
55-60	x		
60-65			

Chapter 4

Zooarchaeological Analysis of Small Vertebrates from Stanton's Cave, Arizona

by

John W. Olsen
Department of Anthropology
The University of Arizona, Tucson

and

Stanley J. Olsen
Arizona State Museum
The University of Arizona, Tucson

Introduction

This paper reports on the analysis of small vertebrate remains recovered in excavations in Stanton's Cave, Grand Canyon, Arizona (Ariz C:5:3). Both mammalian and reptilian elements were identified in the faunal assemblage which represents a living community that may be considered characteristic in most respects of the Grand Canyon today.

Archaeological Interpretations Based on Fauna Recovered

As Euler (1978:147-149) reported, only 15 non-figurine artifacts were recovered during the two seasons of excavation at Stanton's Cave. While a total of perhaps 165 split-twig figurines have been brought to light in Stanton's Cave (only 84 were retrieved as a result of scientific excavation, the rest being taken by hikers, boaters, and pothunters), the rest of the site's artifactual assemblage consists of three small *Olivella* shell beads, one quartzite end scraper, a small one-hand mano, 12 fragments of cordage, and one wooden spindle whorl.

Given this paucity of cultural remains, it isn't surprising that none of the bone we examined appeared to have been culturally modified in any way, either for use as a tool, or as a result of food processing. While some of the fragments recovered exhibited rodent gnaw marks, none displayed the depressed fractures often characteristic of human butchering practices. The palaeoecological data to be derived from this faunal assemblage seem more productive than an attempt to formulate cultural interpretations based on the material recovered to date. Most of the species recovered, including the most abundant ones (*Neotoma* sp. indet. and *Peromyscus* sp. indet.) are almost certainly natural inhabitants of the cave and are not related to human activity within the grotto.

Intrusive rodent remains are a common problem in the interpretation of archaeological faunal remains. It is unfortunate that such a great quantity of the osteological remains recovered in Stanton's Cave may have nothing whatsoever to do with human habitation in the area. Aside from the genera mentioned above, the remains of bats (*Eptesicus* cf. *E. fuscus* and *Myotis* sp. indet.), ringtail (*Bassariscus astutus*), and two genera of squirrels (*Sciurus* and *Spermophilus*) all represent forms which commonly inhabit rocky grottoes such as Stanton's Cave irregardless of human presence.

Site Taphonomy

As Efremov (1940) and Behrensmeyer (1975) have pointed out, zooarchaeological interpretations are to a great extent influenced by the quantity and quality of osteological remains recovered. In the case of the material excavated from Stanton's Cave, we feel a discussion of taphonomic factors and their probable effect on the faunal assemblage is warranted.

Being a cave deposit, the Stanton's Cave death assemblage has been insulated from many of the taphonomic processes that commonly influence the composition of faunal collections from open-air sites. Percolating ground water, rainfall, floods, and snow-melt probably played only a small role in sorting and dispersing the osteological remains recovered in the excavations, although such factors may have had a differential effect on bone depending upon the specific location in the cave or the talus where the bone was located. Periodic fluctuation of moisture content in soils is a frequent cause of damage to archaeological faunal assemblages. Euler (1978: 156-158) has demonstrated it is probable that Stanton's Cave had been subjected to inundation by the Colorado River, creating bedded lenses of driftwood, and it is certain that such flooding could have caused the destruction or displacement of significant portions of the site's original death assemblage.

Obviously, osteological remains associated with the more recent strata were probably not subjected to such disturbance as the Colorado River has not been known to reach flood stage as high as Stanton's Cave (some 44 meters above present water level) in historic times. However, it seems logical that the marked decrease in the quantity of osteological remains toward the earlier end of the Stanton's Cave sequence can at least in part be explained by this extreme climatological event. In all probability, the factor which had the most significant effect on the accumulation of bone in Stanton's Cave was the action of wood rats (*Neotoma* sp. indet.). The behavior pattern whereby this animal builds large middens of cacti, sticks, and assorted small objects is well documented (e.g., Hoffmeister 1971:136; Hall and Kelson 1959:680). The presence of *Neotoma* nests within Stanton's Cave is apparently not a recent phenomenon, although as will be discussed later, remains of this animal become much more abundant in levels that we assume to be post-Pleistocene in age.

Much of the small rodent bone we examined was fragmented in a manner similar to the remains of owl pellets, and as osteological elements of the great horned owl (*Bubo virginianus*) have been recovered in the grotto (Euler 1978:153), the possibility that owls may be at least partially responsible for the accumulation of the smaller fragments of bone seems apparent.

Lithostatic damage may account for much of the bone fragmentation, especially in the deeper strata. Rock fall from the cavern's roof undoubtedly contributed to the destruction of osteological elements on the cave floor.

In summary, it is apparent a broad spectrum of taphonomic factors have affected the composition of the death assemblage recovered from Stanton's Cave. In all probability, the majority of bone displacement may be attributed to bioturbation, primarily on the part of the wood rat (*Neotoma* sp. indet.). Bone destruction appears to have resulted from a combination of factors including lithostatic pressure, hydraulic abrasion and certainly the actions of carnivores and rodents.

The Stanton's Cave fossil assemblage is not unusual in regard to the degree of bone preservation, but taphonomic data do not yield evidence of any relationship between the osteological elements recovered in the grotto and possible human activity in the area.

Mammalian Remains Recovered in Stanton's Cave

Although the analysis of large mammalian osteological material is the subject of another report (Harington, this volume), bones of immature and possibly foetal artiodactyls have been treated in the present paper.

A total of 1917 small mammalian bone fragments were recovered by the Stanton's Cave excavation team, of which 709 (36.95%) were identifiable to at least the level of order. The bulk of the collection (1208 specimens, or 63.05% of the total) consisted of remains too fragmentary to allow useful taxonomic assignment.

The Stanton's Cave faunal assemblage affords no surprises in terms of species present with the exception of remains of the muskrat *(Ondatra zibethicus)*. This animal, no longer included in checklists of Grand Canyon mammalian fauna (Hoffmeister 1971), was positively identified on the basis of a mummified hind limb, complete cranium and left mandible (both of which were burned) and other elements representing at least two individuals. All other mammalian remains recovered represent species extant in the area today.

Discussion of Mammalian Fauna in Stratigraphic Context

Order: Insectivora

Only one insectivore, the desert shrew *(Notiosorex* cf. *N. crawfordi)* was recovered in the Stanton's Cave excavations. Represented by a single mandible with dentition from the surface of the cave floor, little useful interpretation can be formulated on the basis of this solitary element. Hoffmeister (1971:47-48) reports the desert shrew generally inhabits areas of the Grand Canyon below the piñon-juniper zone characterized by sagebrush or cactus. Thus, while other insectivores found in the Grand Canyon (such as Merriam's shrew, *Sorex merriami*, and the dwarf shrew, *Sorex nanus;* Ruffner and Carothers 1975:154) prefer the coniferous forests of the Canyon's rims, the presence of the desert shrew is indicative of an arid environmental niche.

Order: Chiroptera

Two genera contained within this order were identified in the Stanton's Cave assemblage. Remains of the big brown bat *(Eptesicus* cf. *E. fuscus)* were recovered in the 10-20 cm level as well as in a stratigraphic horizon in another part of the cave at a depth of 20-25 cm below the cavern floor. The total remains only represent a minimum of one individual, so detailed interpretation is not possible.

The genus *Myotis* (mouse-eared bats) was identified from strata at a depth of 15-40 cm — perhaps at least 6000 years old (Euler 1978:159).

Hoffmeister (1971:57) reports that while the big brown bat generally inhabits the coniferous forests of both the North and South rims, it may in winter move to the Canyon's bottom to escape the cold. Such exploitation of broadly divergent ecozones obviously limited the animal's utility as an indicator of past climatic conditions.

It seems curious that so few chiropteran remains were found in Stanton's Cave considering the secretive habits of these animals. Destruction of the bats' characteristically fragile bone as a result of any one of a number of taphonomic factors could account for the paucity of chiropteran skeletal elements recovered. Excavation sample bias, such as as the size of the screen used for processing excavated material from various parts of the site may also be responsible for the lack of quantities of bat bones.

Order: Lagomorpha

The remains of two common lagomorphs, the black-tailed jackrabbit *(Lepus californicus)* and the desert cottontail *(Sylvilagus* cf. *S. audubonii)* were identified. The black-tailed jackrabbit is represented by only five fragments indicating a minimum of two individuals. Most of the *Lepus* material was recovered above the 5 cm level, but two fragments of bone are associated with the 15-40 cm horizon. These latter remains fall below Euler's hypothetical 25 to 30 cm Pleistocene/Holocene boundary, so the possibility exists that the Stanton's Cave area provided suitable habitat for *Lepus* as far back as the late Pleistocene.

Eleven fragments of *Sylvilagus* bone were recovered representing at least three individuals. Lagomorphs occurred stratigraphically in levels from the surface down to the 40-65 cm level in the case of *Sylvilagus.* Although representation in the strata is sporadic, the depth to which the remains of lagomorphs have been found is well below the Pleistocene to Holocene transition zone. Hoffmeister (1971:88-89) reports that *Sylvilagus* is only found on the Canyon's South Rim currently, so the occurrence of the genus at Stanton's Cave is of interest from a zoogeographic standpoint as well.

Order: Rodentia

The rodents are the best represented order in the faunal assemblage. Seven genera incorporating a wide range of body sizes, habitat preferences and behavioral patterns comprise this order at Stanton's Cave.

The following genera and species of rodents have been identified: squirrel *(Sciurus* sp. indet.), rock squirrel *(Spermophilus* sp. indet.), beaver *(Castor canadensis)*, wood rat *(Neotoma* sp. indet.), white-footed mouse *(Permyscus* sp. indet.), muskrat *(Ondatra zibethicus)*, and porcupine *(Erethizon dorsatum)*.

By far the most numerous genus, *Neotoma,* is represented by a total of 464 bone fragments from nearly all excavated levels of the site. The importance of wood rat middens in palaeoecological studies has been discussed elsewhere (Wells and Berger 1967), yet it should be emphasized here that in the case of Stanton's Cave, the presence of *Neotoma* over a long period of time has proved to be a mixed blessing. While it is true much valuable palaeoenvironmental data can be gleaned from wood rat nests, it is also unfortunate that the collecting and hoarding activities of this rodent have virtually negated any possible temporal correlation of faunal remains recovered on the present surface of the cave or in the middens themselves. For example, a *Teratornis merriami* humerus retrieved from an unstratified wood rat midden in Stanton's Cave (Euler 1978:159) was radiocarbon dated to 15,320 ± 240 B.P. (University of Arizona Sample A1238). The possibility of intrusive Pleistocene fossils in much later stratigraphic contexts as a result of wood rat midden formation is a substantial problem in the Stanton's Cave assemblage.

Hall and Kelson (1959:680) report, ". . . wood rats, as a genus, occur from low, hot, dry deserts or humid jungles to rocky slopes above timberline." In the absence of complete series of dentition, identification of the *Neotoma* remains to species level is not possible. Therefore, it cannot be positively asserted whether or not more than one species of *Neotoma* is present in the Stanton's Cave faunal assemblage which might allow more precise interpretation of paleoclimatic conditions based upon the presence or relative frequency of a particular species. Even though a fairly large sample of wood rat remains was recovered (over 65% of the identifiable bone), representing at least 39 individuals, meaningful interpretation or temporal correlation of the bones is limited. The burrowing habits of this rodent make the possibility of its intrusion into earlier archaeological strata apparent.

Although little can be derived from the relative frequency of *Neotoma* remains by stratigraphic level, there seems to be a trend toward increasing numbers of the rodent in the upper levels of the site.

Aside from 20 wood rat fragments recovered in stratigraphic levels below 40 cm, the great abundance of these rodents begins to appear near the top of the 25-50 cm stratigraphic unit. It is of especial interest to note that a distinct hiatus exists in the Stanton's Cave stratigraphic profile between these two occurrences of bone in which *no* identifiable fragments, including *Neotoma* were recovered. Wood rat remains are especially abundant in the 5-30 cm unit, where 102 fragments of bone were recovered. Because radiocarbon dates based on organic material from these levels have produced chronometric dates ranging from 2450 ± 80 B.P. to 13,770 ± 500 B.P., it is impossible to pin down exactly when this increase in *Neotoma* population began, but it appears that the phenomenon is linked with post-Pleistocene desiccation of the Stanton's Cave area.

The second-most numerous rodent in the collection analyzed was the white-footed mouse (*Peromyscus* sp. indet.). Ninety fragments of bone representing at least 15 individuals were recovered from eight separate stratigraphic levels. It is of importance to note that all but one fragment (from the 45-70 cm unit) were found between 50 cm and the present surface of the cave floor. This relative frequency of *Peromyscus* bone correlates well with the differential abundance of *Neotoma* remains discussed above.

Of the five species of *Peromyscus* occurring in the Grand Canyon today (Hoffmeister 1971:35-36; Hall and Kelson 1959:600-658; Bailey 1935) only the piñon mouse (*Peromyscus truei*) require environmental conditions not extant in the immediate vicinity of Stanton's Cave at the present time. The peromyscine mice are noted for their adaptability to a wide variety of environmental niches, yet each species has definite habitat preferences (Holbrook 1978:7-9).

As *Peromyscus* remains are not identifiable to the species level on the basis of post-cranial material, the possible range of environmental conditions represented by their remains at Stanton's Cave is very great.

If Euler is correct in placing the Pleistocene/Holocene boundary at the 25-30 cm level in the cave's stratigraphic sequence, then it is undoubtedly of relevance that *Peromyscus* remains become particularly abundant at this same level. If, as other sources indicate, the Southwest experienced a trend toward desiccation at the end of the Pleistocene, one might expect the *Peromyscus* material from Stanton's Cave to be those of *P. eremicus* or possibly *P. crinitus,* both of which inhabit xeric environments. On the basis of overall size alone, the *Peromyscus* remains recovered resemble these species, but in the absence of well-preserved dental and cranial material to compare, this constitutes pure speculation.

Rodents recovered that indicate a riparian habitat in the vicinity of Stanton's Cave include the beaver (*Castor canadensis*) and muskrat (*Ondatra zibethicus*). Hoffmeister (1971:122-124) reports that beavers have been known historically in the Grand Canyon since 1826. Known to inhabit tributaries of the Colorado River and sighted on the river itself at a number of places, it is not surprising that *Castor* remains were found in the cave, particularly in light of the proximity of a secondary source of permanent water — Vaseys Paradise. Only five fragments of beaver bone, representing a minimum of one individual were recovered from surface collections in the cave.

The muskrat (*Ondatra zibethicus*) is a surprising find as it is not known to inhabit the Grand Canyon today. While at least two subspecies of the muskrat (*O. z. goldmani* and *O. z. osoyoosensis*) have been reported from southern Utah and western Arizona, their zoogeographic range is extrapolated by Hall and Kelson (1959:756) to include extreme northern Arizona, possibly as far south as the Grand Canyon.

A total of four elements, including a nearly complete burned cranium, were recovered which represent at least two individuals. While the majority of the muskrat bones were found on the present surface of the cave floor, one element was found at a depth of at least 5 cm. Of all the mammalian forms reported on in this paper, the muskrat is the only one not known to inhabit the Grand Canyon vicinity at the present time.

The porcupine (*Erethizon dorsatum*) is reported by Hoffmeister (1971:146-148) as being common on both rims as well as in the interior of the Grand Canyon. Only one fragment of porcupine bone was recovered, and its location in the 5-30 cm unit is not easily interpreted, for as Hoffmeister (1971:147) states, *Erethizon* frequently lives in burrows not unlike a badger *(Taxidea taxus)*, so the possibility of intrusion in this case is apparent.

Order: Carnivora

Carnivores of six genera representing four families (Canidae, Procyonidae, Mustelidae, and Felidae) were recovered in the Stanton's Cave excavations.

Ringtail *(Bassariscus astutus)* remains were by far the most common with a minimum of three individuals represented. Sixty of the 81 *Bassariscus* bone fragments recovered originated on the present surface of the cave floor or in wood rat nests out of their original stratigraphic context. Like the rest of the carnivore element uncovered, the *Bassariscus* material all comes from provenances which are probably post-Pleistocene in age. As habitation preferences of the ringtail include rocky grottoes, the possibility of their being an intrusive element in the Stanton's Cave faunal assemblage must be considered.

The other procyonid identified in the faunal assemblage was the raccoon *(Procyon lotor)*. Only two fragments of *Procyon* bone were found, both from the present surface of the cave floor.

Since riparian communities such as those found on the banks of the Colorado River and at Vaseys Paradise near Stanton's Cave provide an ideal habitat for raccoons, the presence of this animal is not surprising. The subspecies of *P. lotor* currently residing in the area of the Grand Canyon is *P. lotor pallidus* (Hall and Kelson 1959:885; Cockrum 1960:229); however on the basis of the scant osteological material recovered, no assignment to the subspecific level can be made.

Canids recovered include both the coyote *(Canis latrans)* of which two elements were identified, and the grey fox *(Urocyon cinereoargenteus)*, which was represented by nine fragments. Both animals were represented only in post-Pleistocene levels above 20 cm in depth.

Six fragments of mountain lion or puma *(Felis concolor)* bone were found on the present surface of the cave and at a depth of 10-20 cm. As there is usually a proportional relationship between puma populations and that of the mule deer *(Odocoileus hemionus)*, the relative scarcity of deer within the confines of the Canyon itself may indicate puma were never especially common below the rims (see Hoffmeister 1971:84).

The only mustelid identified from the Stanton's Cave faunal assemblage is the river otter *(Lutra canadensis)*. While Hoffmeister (1971:72) states that less than a dozen sightings of *Lutra* have been reported from the Grand Canyon in historic times, Mearns (1891) includes the Canyon within its zoogeographic range. The fact that only one fragment of *Lutra* bone was recovered precludes detailed interpretation. However, as the fragment was located at a depth of 25-50 cm, this find may possibly indicate a riparian habitat at Stanton's Cave in the late Pleistocene, perhaps before 13,000 B.P.

Order: Artiodactyla

Although the larger mammals from Stanton's Cave have been reported on elsewhere (Harington, this volume), remains of several immature artiodactyls were included in the smaller faunal assemblage and will be discussed here. Osteological elements of both cervids and bovids were recovered.

The mule deer *(Odocoileus hemionus)* has been positively identified on the basis of mandibular, vertebral, and other elements. All *Odocoileus* remains occurred above the 5 cm level so it is assumed their presence in the cave is a recent phenomenon. The identification of very immature or possibly foetal skeletal elements may indicate their introduction into the cavern by human or predatory agencies.

The bighorn or mountain sheep *(Ovis canadensis)* is represented by a variety of elements including two partial horn cores. Occurring in post-Pleistocene strata above 20 cm, the bighorn is probably responsible for filling the ecological niche vacated by *O. harringtoni* at the close of the Pleistocene (see Harington, this volume).

Hoffmeister (1971:155) reports that mountain sheep, ". . . are frequently seen by river boats as they water at the river's edge." As with so many of the mammalian species identified in the Stanton's Cave faunal collection, the combination of the proximity of the Colorado River's riparian habitat and Vaseys Paradise might have acted as a strong inducement for the bighorn to frequent the area.

Conclusions Based on the Mammalian Fauna

In summary, a few broad generalizations can be offered regarding palaeoecological conditions at Stanton's Cave. While the general paucity of identifiable faunal remains recovered does not allow detailed interpretation, it seems apparent that substantial changes in species composition and population frequencies occurred at about the same place in the stratigraphic sequence that Euler feels delineates the Pleistocene/Holocene boundary.

While the species present in the Stanton's Cave death assemblage present few surprises, they generally seem to indicate a riparian habitat bounded by a mosaic of rocky outcroppings with occasional piñon-juniper stands. While this palaeoenvironmental picture may not remain constant throughout the Stanton's Cave sequence, the small quantity of faunal remains recovered below the present surface of the cave does not permit a detailed study of diachronic change.

The Herpetofauna of Stanton's Cave

Reptile remains recovered from the excavations occurred less frequently than might be expected since this class is a common inhabitant of the area today. We would not hazard a guess as to the possible palaeoecological implications of the mere scarcity of these forms.

All of the lizards recovered are regarded as diurnal forms (Gehlbach 1966). The chuckwalla *(Sauromalus obesus)* inhabits rocky, shrub desert in the Canyon below 1300 meters. The collared lizard *(Crotaphytus collaris)* also frequents rocky, shrub desert, but also has expanded its range into evergreen woodland areas on both rims of the Canyon below 2200 meters. The fence lizard *(Sceloporus undulatus)* inhabits evergreen woodlands and shrub desert below 2500 meters (Gehlbach 1966; Stebbins 1966). All of these lizards are currently abundant within the Canyon.

The kingsnake *(Lampropeltis getulus)* inhabits shrub desert and deciduous woodland in the Canyon below 1300 meters (Gehlbach 1966). It is generally regarded as being noctural and is frequently encountered in the Canyon today. Snake bones were the most numerous of the reptile elements recovered, but the problem of determining whether or not they are intrusive is a great one. *Lampropeltis getulus* was represented by vertebrae, both isolated and articulated, which occurred in surface collections from the cave floor as well as in the 15-40 cm unit in the excavations. Several vertebrae were retrieved from *Neotoma* middens as were mandibles and vertebrae of the chuckwalla *(Sauromalus obesus)*.

Unfortunately, the few reptilian osteological remains recovered from Stanton's Cave (a total of 36 fragments) do not permit accurate palaeoclimatic reconstruction or even whether their presence in the cave is the result of human actions, predatory hunters, or simple intrusion on the part of the reptiles themselves. However, with these limitations in mind, several conclusions can be offered. First, all of the reptilian species represented occur in either evergreen woodland, deciduous woodland, or rocky, shrub desert within an altitudinal range from about 2500 meters down to the present level of the Colorado River. Not enough reptile remains were found to allow comparison of species frequency through time, but it seems apparent that the herpetofauna of Stanton's Cave remained stable in terms of species diversity at least down to the 15 cm level of the excavation, or about 6000 years ago (Euler 1978:159).

Like rodents, many reptiles are noted for their burrowing habits. In particular the chuckwalla *(Sauromalus obesus)* which inhabits rock crevices, and the kingsnake *(Lampropeltis getulus)* which also frequents rocky areas and rodent burrows, are likely to be found in an ecological niche such as Stanton's Cave regardless of the presence or absence of human occupants. The smaller lizards *(Crotaphytus collaris* and *Sceloporus undulatus)* probably in no way reflect subsistence activities of the region's human inhabitants.

The absence of amphibian remains is puzzling, especially in light of the proximity of the spring at Vaseys Paradise, an ideal breeding ground for such forms. Taphonomic factors most likely account for their absence due to the fragile structure of amphibian bone. Euler (1978:147) reports, "All excavated material was carefully screened, the 5-centimeter levels through a window screen and the others through a 6-millimeter mesh." While window screen mesh (about 1 millimeter) is ordinarily small enough to permit recovery of even the smallest osteological elements, most amphibian bones will easily pass through 6 millimeter mesh. Therefore, sampling may in part account for both the paucity of reptile remains and the lack of amphibian elements recovered in Stanton's Cave.

Table 4-1. Taxonomic list of small mammalian fauna present.

Order: Insectivora
- Family: Soricidae
  - *Notiosorex* cf. *N. crawfordi*, desert shrew

Order: Chiroptera
- Family: Vespertilionidae
  - *Eptesicus* cf. *E. fuscus*, big brown bat
  - *Myotis* sp. indet., mouse-eared bat

Order: Lagomorpha
- Family: Leporidae
  - *Lepus californicus*, black-tailed jackrabbit
  - *Sylvilagus* cf. *S. audubonii*, desert cottontail

Order: Rodentia
- Family: Sciuridae
  - *Sciurus* sp. indet., squirrel
  - *Spermophilus* sp. indet., rock squirrel
- Family: Castoridae
  - *Castor canadensis*, beaver
- Family: Cricetidae
  - *Neotoma* sp. indet., wood rat
  - *Peromyscus* sp. indet., white-footed mouse
  - *Ondatra zibethicus*, muskrat
- Family: Erethizontidae
  - *Erethizon dorsatum*, porcupine

Order: Carnivora
- Family: Canidae
  - *Canis latrans*, coyote
  - *Urocyon cinereoargenteus*, grey fox
- Family: Procyonidae
  - *Bassariscus astutus*, ringtail
  - *Procyon lotor*, raccoon
- Family: Mustelidae
  - *Lutra canadensis*, river otter
- Family: Felidae
  - *Felis concolor*, mountain lion or puma

Order: Artiodactyla
- Family: Cervidae
  - *Odocoileus hemionus*, mule deer or black-tailed deer
- Family: Bovidae
  - *Ovis canadensis*, bighorn or mountain sheep

Table 4-2. Mammalian fauna: quantitative data.

Taxon	Total Bones	MNI*	#Burned	%Burned	% of Total Bone in Site	% of Identifiable Bone
Notiosorex cf. *N. crawfordi*	1	1			.05	.14
Eptesicus cf. *E. fuscus*	4	1			.21	.56
Myotis sp. indet.	1	1			.05	.14
Lepus californicus	5	2	1	20	.26	.71
Sylvilagus cf. *S. audubonii*	11	3			.57	1.50
Sciurus sp. indet.	2	1			.10	.28
Spermophilus sp. indet.	3	1			.16	.42
Castor canadensis	5	1			.26	.71
Neotoma sp. indet.	464	39			24.20	65.00
Peromyscus sp. indet.	90	15			4.70	13.00
Ondrata zibethicus	4	2	3	75	.21	.56
Erethizon dorsatum	1	1			.05	.14
Canis latrans	2	1			.10	.28
Urocyon cinereoargenteus	9	1	1	11	.47	1.28
Bassariscus astutus	81	3	19	23.4	4.20	11.40
Procyon lotor	2	1	1	50	.10	.28
Lutra canadensis	1	1			.05	.14
Felis concolor	6	1			.31	.85
Odocoileus hemionus	3	1			.16	.42
Ovis canadensis	5	2	1	20	.26	.71
Artiodactyl gen. et sp. indet.	9	3	4	44	.47	1.30
Unidentifiable	1208	—	—		63.05	—
Totals	1917	81	31		99.99	100.00

*MNI = Minimum Number of Individuals

Table 4-3. Occurrence of mammalian remains by horizontal provenance.

Surface and Wood Rat Nests
Talus Outside of Fence
- *Lepus californicus*
- *Neotoma* sp. indet.

Surface Within Cave
- *Notiosorex* cf. *N. crawfordi*
- *Sylvilagus* cf. *S. audubonii*
- *Spermophilus* sp. indet.
- *Neotoma* sp. indet.
- *Peromyscus* sp. indet.
- *Canis latrans*
- *Felis concolor*
- *Odocoileus hemionus*
- *Ovis canadensis*
- Artiodactyl, gen. et sp. indet.

Collected from Wood Rat Nests
- *Lepus californicus*
- *Sylvilagus* cf. *S. audubonii*
- *Sciurus* sp. indet.
- *Neotoma* sp. indet.
- *Peromyscus* sp. indet.
- *Ondatra zibethicus*
- *Castor canadensis*
- *Urocyon cinereoargenteus*
- *Bassariscus astutus*
- *Procyon lotor*
- *Felis concolor*
- *Ovis canadensis*

East-West Test Trench
Grid AA
- *Neotoma* sp. indet.
- *Peromyscus* sp. indet.

Grid BB
- *Neotoma* sp. indet.
- *Peromyscus* sp. indet.

Grid CC
- *Neotoma* sp. indet.
- *Peromyscus* sp. indet.
- *Odocoileus* cf. *O. hemionus*

Grid DD
- *Neotoma* sp. indet.
- *Peromyscus* sp. indet.
- *Bassariscus astutus*

Grid EE
- *Neotoma* sp. indet.
- *Peromyscus* sp. indet.

Grid FF
- *Myotis* sp. indet.
- *Neotoma* sp. indet.
- *Peromyscus* sp. indet.
- *Bassariscus astutus*

Grid GG
- *Neotoma* sp. indet.
- *Peromyscus* sp. indet.
- *Bassariscus astutus*

Grid HH
- *Neotoma* sp. indet.
- *Peromyscus* sp. indet.
- *Bassariscus astutus*

North-South Test Trench
Grid G-19
- *Lepus californicus*
- *Sylvilagus* cf. *S. audubonii*
- *Neotoma* sp. indet.
- *Peromyscus* sp. indet.
- *Ovis canadensis*

Grid H-19
- *Neotoma* sp. indet.
- *Peromyscus* sp. indet.
- *Lutra canadensis*

Grid I-19
- *Sylvilagus* cf. *S. audubonii*
- *Spermophilus* sp. indet.
- *Neotoma* sp. indet.
- *Peromyscus* sp. indet.
- Artiodactyl, gen. et sp. indet.

Grid J-19
- *Sciurus* sp. indet.
- *Neotoma* sp. indet.
- *Ondatra zibethicus*
- *Erethizon dorsatum*

Grid K-19
- *Neotoma* sp. indet.
- *Ovis canadensis*
- Artiodactyl, gen. et sp. indet.

Grid L-19
- *Sylvilagus* cf. *S. audubonii*

Grid M-19
- *Neotoma* sp. indet.

Grid O-19
- *Neotoma* sp. indet.
- *Peromyscus* sp. indet.
- Artiodactyl, gen. et sp. indet.

Grid P-19
- *Eptesicus* cf. *E. fuscus*
- *Neotoma* sp. indet.

Grid Q-19
- Small mammal, gen. et. sp. indet.

Table 4-4. Taxonomic list of herpetofauna present.

Order: Squamata
  Suborder: Lacertilia
    Family: Iguanidae
      Sauromalus obesus, chuckwalla
      Crotaphytus collaris, collared lizard
      Sceloporus undulatus, fence lizard

  Suborder: Serpentes
    Family: Colubridae
      Lampropeltis getulus, kingsnake

Stratigraphic occurrence of reptile remains.

Sauromalus obesus, chuckwalla - total: 7 bones
  Wood rat nest adjacent to FS-14: left mandible with dentition
  Wood rat nest near FS-3: 6 vertebrae
Crotaphytus collaris, collared lizard - total: 1 bone
  Grid FF: 10-15 cm: left maxilla with dentition
Sceloporus undulatus, fence lizard - total: 1 bone
  H-19: 10-15 cm: right mandible with dentition
Lampropeltis getulus, kingsnake - total: 27 bones
  Antechamber 1, surface: 12 articulated vertebrae
  Grid CC: 20-25 cm: 1 vertebra
  Grid FF: 20-25 cm: 1 vertebra
  G-19: 10-15 cm: 1 vertebra
  G-19: 15-40 cm: 1 vertebra
  H-19: 10-15 cm: 4 vertebrae
  I-19: 20-25 cm: 1 vertebra
  Q-19: 5-10 cm: 2 vertebrae
  West of FS-2: 1 vertebra
  Wood rat nest near FS-3 and FS-5: 3 vertebrae

TOTAL: 36 elements

References

Bailey, Vernon

1935 Mammals of the Grand Canyon region. *Natural History Bulletin* 1:1-42. Grand Canyon: Grand Canyon Natural History Association.

Behrensmeyer, Anna K.

1975 The taphonomy and paleoecology of Plio-Pleistocene vertebrate assemblages east of Lake Rudolf, Kenya. *Bulletin of the Museum of Comparative Zoology* 146:10:474-578. Cambridge, Massachusetts: Harvard University.

Cockrum, E. Lendell

1960 *The Recent mammals of Arizona: Their taxonomy and distribution.* Tucson: University of Arizona Press.

Efremov, J.A.

1940 Taphonomy: A new branch of palaeontology. *Pan-American Geologist,* Volume 74, 2:81-93.

Euler, Robert C.

1978 Archeological and Paleobiological Studies at Stanton's Cave, Grand Canyon National Park, Arizona — A Report of Progress. *National Geographic Society Research Reports, 1969 Projects* 141-162.

Gehlbach, Frederick R.

1966 *Grand Canyon amphibians and reptiles: Field checklist.* Flagstaff: Northland Press.

Hall, E. Raymond and Keith R. Kelson

1959 *The mammals of North America,* two volumes. New York: The Ronald Press.

Hoffmeister, D.F.

1955 Mammals new to Grand Canyon National Park, Arizona. *Plateau,* Volume 28, 1:1-7.

1971 *Mammals of Grand Canyon.* Urbana: University of Illinois Press.

Hoffmeister, D.F., and F.E. Durham

1971 Mammals of the Arizona Strip including Grand Canyon National Monument. *Museum of Northern Arizona Technical Series,* No. 11, Flagstaff.

Holbrook, Sally J.

1978 *Paleoecology of Grasshopper Pueblo, Arizona.* Unpublished manuscript submitted to National Geographic Society for publication in *Research Reports.*

Lowe, Charles H., editor

1964 *The vertebrates of Arizona.* Tucson: University of Arizona Press.

Mearns, Edgar A.

1891 Notes on the otter *(Lutra canadensis)* and skunks (Genera *Spilogale* and *Mephitis)* of Arizona. *Bulletin of the American Museum of Natural History* 3:252-262. New York.

Ruffner, G.A., and S. W. Carothers

1975 Recent notes on the distribution of some mammals of the Grand Canyon region. *Plateau* 47:4:154-160.

Stebbins, Robert C.

1954 *Amphibians and reptiles of Western North America.* New York: McGraw-Hill Book Company.

1966 *A field guide to Western reptiles and amphibians.* Peterson Field Guide Series. Boston: Houghton Mifflin Company.

Wells, P.V., and Rainer Berger

1967 Later Pleistocene history of coniferous woodland in the Mohave Desert. *Science* 155:1640-1647.

Wright, G.M., J.S. Dixon, and B.H. Thompson

1933 Fauna of the national parks of the United States: A preliminary survey of faunal relations in the National Parks. *U.S. Department of the Interior Fauna Series,* Volume 1, Number 4. Washington, D.C.

Chapter 5

Fish Remains from Stanton's Cave, Grand Canyon of the Colorado, Arizona, With Notes on the Taxonomy of *Gila cypha*

by

Robert Rush Miller and Gerald R. Smith

Museum of Zoology and Museum of Paleontology
The University of Michigan, Ann Arbor

Stanton's Cave is 51 river kilometers below Lees Ferry, Coconino County, Arizona, in the inner gorge of Marble Canyon, and lies at an elevation of 927 meters. Here the river makes an abrupt S-bend some 50 river kilometers north of the mouth of the Little Colorado River. The entrance to the cave lies 44 meters above the river and measures about 6.1 meters wide and 9.1 meters high; passageways extend almost 61 meters into the limestone bluff. The site is designated Ariz. C:5:3 in the archeological survey records of Grand Canyon National Park.

The fish remains recovered from Stanton's Cave by Robert C. Euler (Table 5-1) represent nearly all of the species known to be native to this part of the Colorado River: the humpback chub, *Gila cypha* Miller; bonytail, *Gila elegans* Baird and Girard; and Colorado squawfish, *Ptychocheilus lucius* Girard; of the Cyprinidae (minnows); the flannelmouth sucker, *Catostomus latipinnis* Baird and Girard, and bluehead sucker, *Catostomus discobolus* Cope, of the Catostomidae (suckers). Among the large riverine fishes, only the razorback sucker, *Xyrauchen texanus* (Abbott), is missing. The roundtail chub, *Gila robusta* Baird and Girard, is rare in the main river, but does occur in the Little Colorado River. The only mainstream records are from near and well above Lees Ferry (Holden and Minckley 1980).

Notes on *Gila cypha*, Humpback chub

Bones of the humpback chub in Stanton's Cave provide evidence for the ancient distinctiveness of this species. This large, bizarre minnow is known to attain a maximum size of 320 mm S.L. Its type locality is in Grand Canyon near the mouth of Bright Angel Creek (Miller 1946). It has since been recorded as far downriver as Boulder Canyon, 39 km below Hoover Dam (from archeological material, Miller 1955), and as far up the Colorado River drainage as Flaming Gorge and Hideout Canyon on the Green River, to just north of the Wyoming border (Smith et al. 1979). We may speculate that it once occurred up the San Juan River at least as far as the Goosenecks, which begin a few kilometers west of Mexican Hat, Utah.

The humpback chub is now known to be present in the Colorado drainage at (1) the canyon region of the lower Yampa and Green rivers in Dinosaur National Monument, (2) Desolation and Gray canyons in the lower Green River, (3) the Black Rocks area in Ruby Canyon, and Westwater Canyon, both on the Colorado River not far above and below the Utah-Colorado state line, and (4) the Grand Canyon region, chiefly the lower 21 km of the Little Colorado River (Behnke and Benson, 1980, and supplementary information).

The controversy over the recognition of the humpback chub as a full species (Miller 1955; Minckley and Deacon 1968; Holden and Stalnaker 1970) has been discussed by Smith et al. 1979, who dealt with morphometric and meristic traits. One potentially valuable organ system, the skeleton, has heretofore remained almost unexplored for systematic characters that might help to resolve the uncertainty over recognizing the humpback chub as a full species. Because the Stanton's Cave material comprises bones only, we have examined the skeletons of big-river members of the genus *Gila* that inhabit the middle and upper parts of the Colorado River basin: *Gila robusta*, *G. elegans*, and *G. cypha*. Our observations are summarized below.

The cave material (Table 5-1) is from: (1) the surface of antechamber 1 (June 17, 1969, R.C. Euler, J. Ware, B. Harrill, collectors), an incomplete neurocranium with Weberian apparatus attached (Figure 5-1D), plus 9 attached vertebrae and 3 ribs; and (2) a pack rat's nest adjacent to FS #14 (June 1969, L. Powers, collector), a left pectoral girdle (Figure 5-3B). All of these remains are non-mineralized; their age may be as great as 4000 years B.P.(Robert C. Euler, personal communication, 1971).

One of the best osteological features that distinguish *Gila cypha* from its relatives is the prominent supraoccipital process (Figure 5-1C,D), which is expanded so that the dorsal border lies above the dorsal profile of the skull (consistent in 5 skulls examined). This enlarged process, evidently developed for muscle attachment associated with the prominent fleshy hump, is also readily seen on radiographs (58 examined, representing materials from widely separated localities) where it is well developed in specimens that have attained 200 mm S.L. In *Gila robusta* (Figure 5-1A) the process is elongate but not elevated, and in *G. elegans* (Figure 5-1B) it is higher than in *robusta* but does not extend above the dorsal profile of the skull.

The angle of the neural and hemal spines to the centra of the posterior caudal vertebrae is a useful feature used to separate *Gila cypha* and its close relatives (Gehlbach and Miller 1961; Minckley and Alger 1968). The vertebral spines of *G. robusta* are least parallel and those of *G. elegans* are most parallel with the centra (Figure 5-2A-D).

The anterior arm of the pectoral girdle is distinctive in *Gila cypha*. A strong dorso-lateral ridge forms a deep sulcus at the base of the posterior (vertical) arm, whereas in both *G. elegans* and *G. robusta* this ridge is weak or obsolete (Figure 5-3A,B).

In summary, the species recovered from Stanton's Cave include, among the minnows (family Cyprinidae), *Gila cypha*, humpback chub, with a weight to nearly 1 kg, locally common; *Gila elegans*, bonytail, with a weight to nearly 1 kg, rare; and *Ptychocheilus lucius*, Colorado squawfish, with a weight to over 20 kg, rare. Among the suckers (family Catostomidae) are *Catostomus latipinnis*, flannelmouth sucker, with a weight to over 1 kg, locally common; and *Catostomus discobolus*, bluehead sucker, with a weight to ca. 5 kg, locally common.

Acknowledgments

We are grateful to Ted Hatch, of Hatch River Expeditions, for facilitating our river survey in 1968 from Lees Ferry to Diamond Creek. In 1975, the first author was sponsored for a similar trip by the National Park Service. In both years, permits to collect were kindly arranged by the Arizona Game and Fish Department and the Superintendent of Grand Canyon National Park. Michael L. Smith determined several of the sucker bones. We thank George Junne for preparing Figures 5-1-3. To Frances H. Miller we are indebted for typing and editing the manuscript.

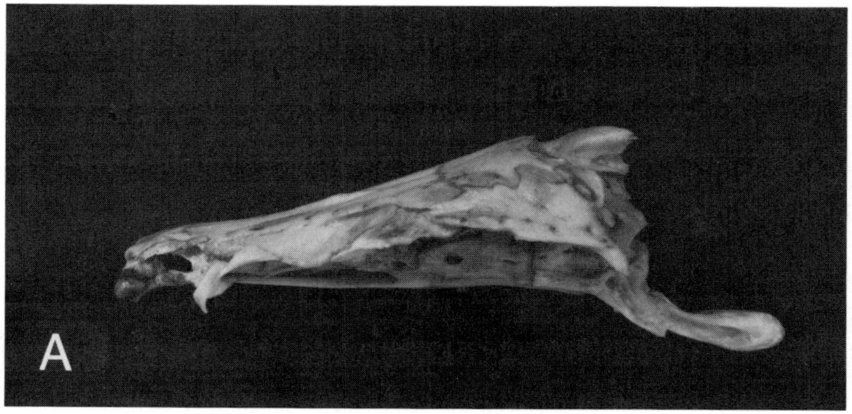

Figure 5-1A. Lateral view of neurocranium of *Gila robusta,* total length 60.3 mm. From a male(?) 281 mm S.L., UMMZ 182546. UTAH:Green River at base Flaming Gorge Dam, IX:7:1962.

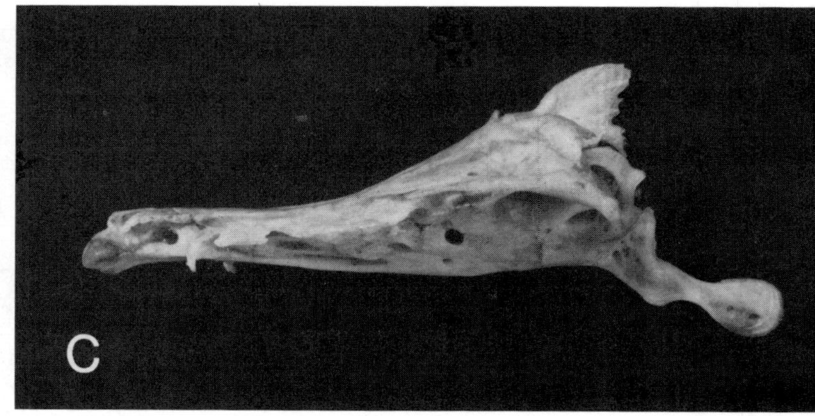

Figure 5-1C. Lateral view of neurocranium of *Gila cypha,* total length 63.0 mm. From a male 295 mm S.L., UMMZ 178667. UTAH: White River, 5.6 km SE of Bonanza, VIII: 16:1966.

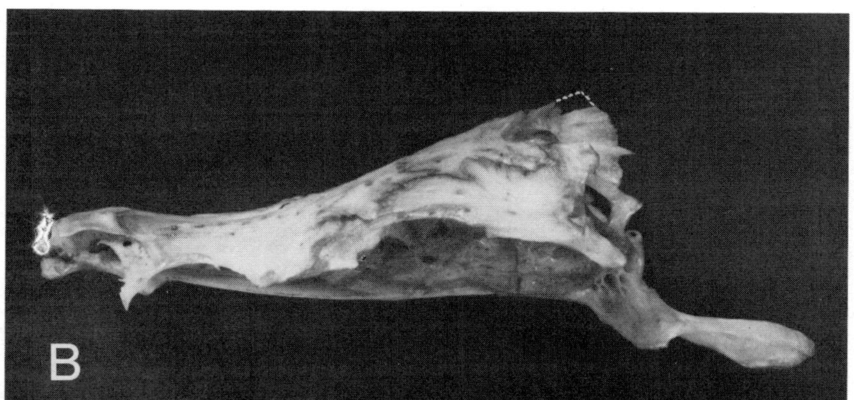

Figure 5-1B. Lateral view of neurocranium of *Gila elegans,* total length 60.5 mm. From a female 330 mm S.L., UMMZ 176972. CALIF:Colorado River, 550 m above Parker Dam, V:17:1959.

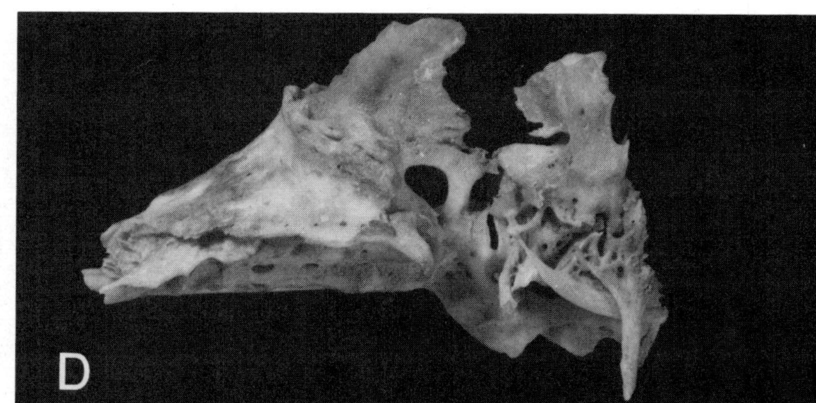

Figure 5-1D. Incomplete neurocranium of *Gila cypha* from Stanton's Cave, total length 54.0 mm. From an individual estimated to be nearly 400 mm S.L.

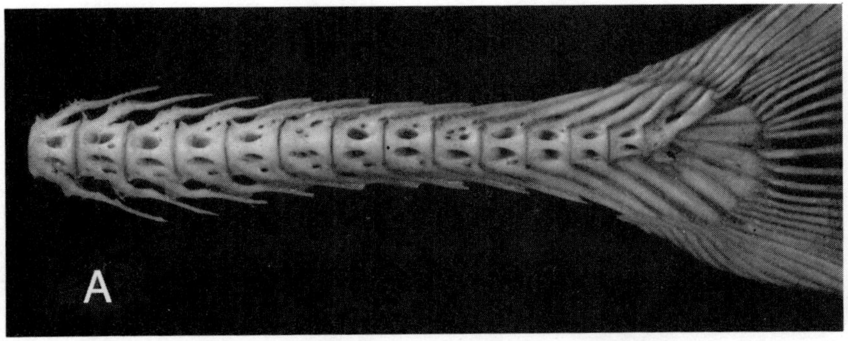

Figure 5-2A. Posterior 14 caudal vertebrae of *Gila robusta*, UMMA 182546 (see Figure 5-1A).

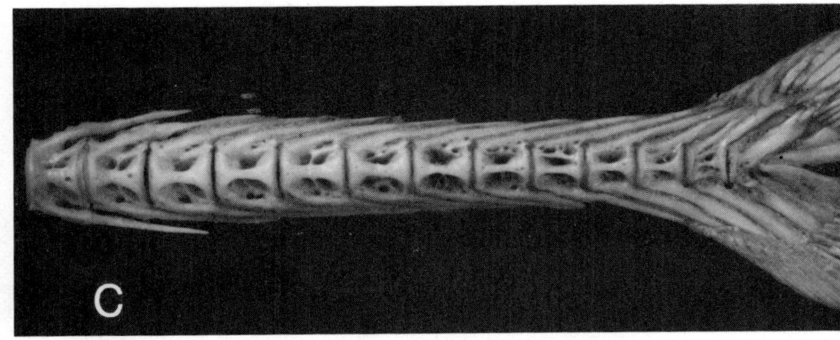

Figure 5-2C. Posterior 13 caudal vertebrae of *Gila elegans*, UMMZ 179581. From a male (?) 331 mm S.L., UTAH:Green River in pool below Flaming Gorge Dam, VII:21:1961.

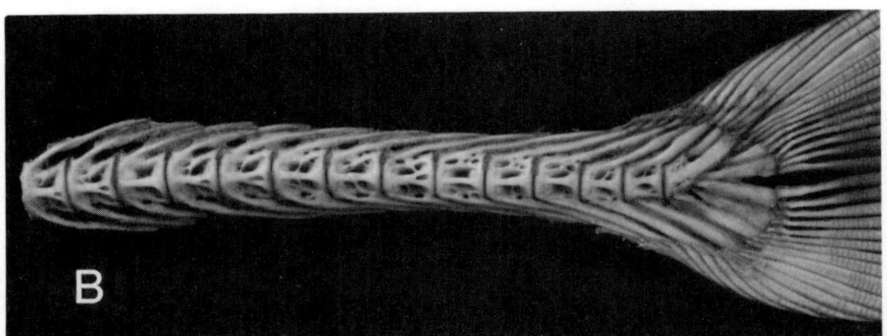

Figure 5-2B. Posterior 14 caudal vertebrae of *Gila cypha*, UMMA 178677 (see Figure 5-1C).

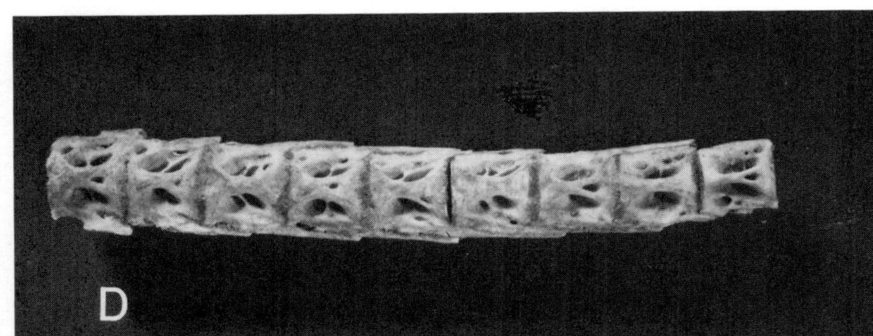

Figure 5-2D. Nine caudal vertebrae of *Gila elegans* from Stanton's Cave, from a specimen estimated to be about 400 mm S.L.

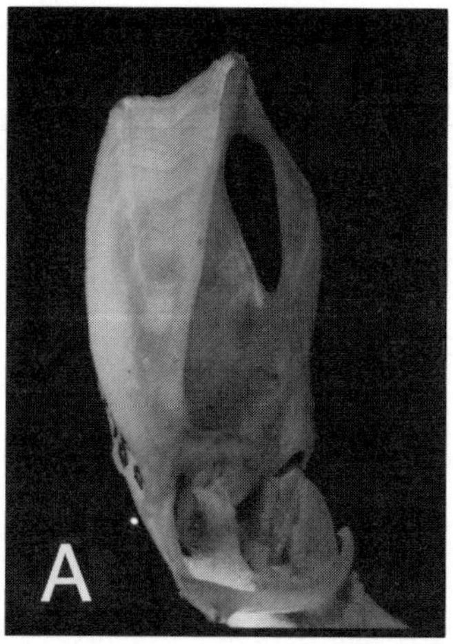

Figure 5-3A. Dorsal view of left pectoral girdle of *Gila elegans*, UMMZ 179581 (see Figure 5-2C).

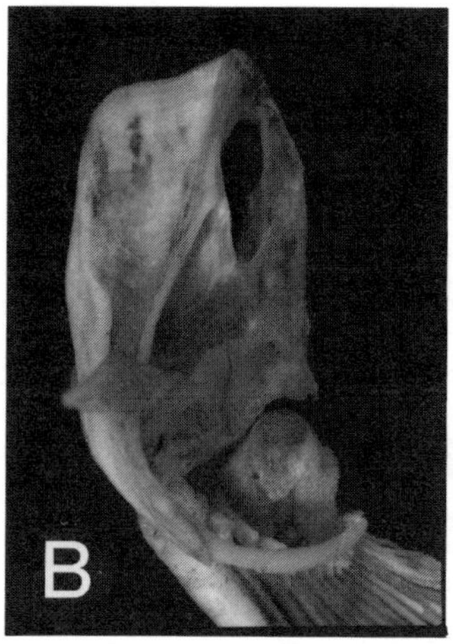

Figure 5-3B. Dorsal view of left pectoral girdle of *Gila cypha* from Stanton's Cave. From a specimen about 300 mm S.L.

Table 5-1. Fish remains from Stanton's Cave, Grand Canyon (UMMZ Acc. 1969-VIII:27, 1970-I:21), Ariz C:5:3.

1. Surface, antechamber 1, R.C. Euler, John Ware III, Bruce Harrill, collectors, June 17, 1969.
   Gila cypha, humpback chub
   1 incomplete neurocranium with Weberian apparatus attached; plus 9 attached vertebrae and 3 ribs. From a fish 300 mm S.L.
   Ptychocheilus lucius, Colorado squawfish
   SC62, right second branchiostegal. From a fish perhaps 1.2-1.5 m long.
2. Pack rat's nest adjacent to FS #14, June 1969, Lawrence Powers.
   Gila cypha
   Left pectoral girdle. From a specimen about 300 mm S.L.

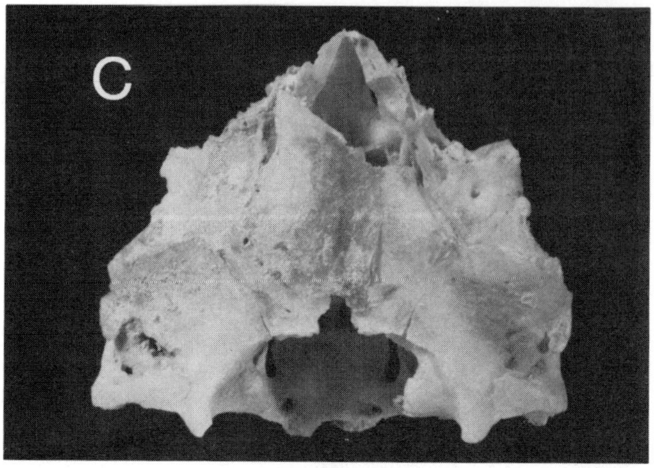

Figure 5-3C. Incomplete neurocrocranium of *Catostomus latipinnis* (anterior is dorsal), total length 33.6 mm, width 37.9 mm, from Stanton's Cave. From a specimen about 400 mm S.L.

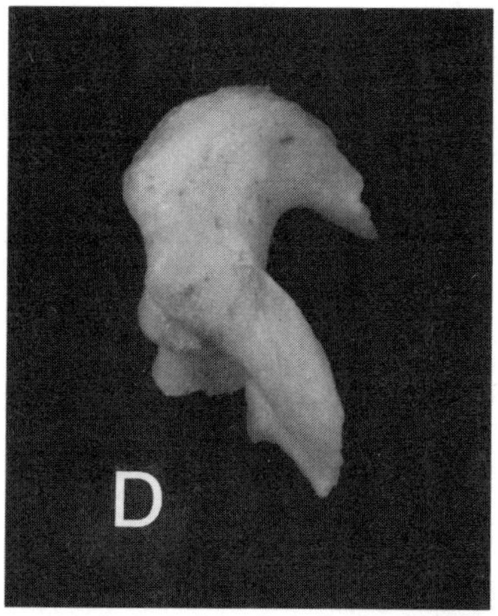

Figure 5-3D. Incomplete left dentary of *Catostomus discobolus* (anterior is dorsal) from Stanton's Cave, total length 12.4 mm. From a specimen about 400 mm S.L.

   Base of left pectoral girdle, 16.6 mm T.L. From a specimen about 400 mm S.L. The pattern of growth-ring density at the edge of the bone indicates that the specimen was taken in the spring of the year.

3. Main room, N-S T. Tr, Grid G-19, 15-40 cm. June 27, 1969, Bruce Harrill.
   Catostomus discobolus, bluehead sucker
   Left dentary (incomplete). From a specimen ca. 315 mm S.L.
   Left premaxilla (broken), 9.0 mm T.L. From a specimen ca. 300 mm S.L.
   Parasphenoid (incomplete), 14.9 mm T.L. From a specimen at least 335 mm S.L.

Base of left pectoral fin-ray, 14.3 mm T.L. From a specimen at least 350 mm S.L.

Catostomus latipinnis, flannelmouth sucker

Neurocranium (incomplete), 33.6 mm T.L. From a specimen ca. 400 mm S.L.

4. Main room, N-S T. Tr, Grid H-19, 25-50 cm. June 25, 1969, John Ware III and Bruce Harrill.
Catostomus discobolus
Left maxilla (broken), 16.2 mm T.L. From a specimen at least 280 mm S.L.

5. Main room, N-S T. Tr, Grid I-19, 25-58 cm. June 26, 1969, John Ware III and Bruce Harrill.
Catostomus discobolus
Dermethmoid (broken), 12.8 mm long, 13.9 mm wide. From a specimen ca. 275 mm S.L.

6. Main room, N-S T. Tr, Grid I-19, 15-20 cm. June 24, 1969, Bruce Harrill and John Ware III.
Catostomus discobolus
Right dentary (incomplete). From a specimen ca. 260 mm S.L.

7. Pack rat's nest near FS #3 and FS #5. June 17, 1969, Robert Page.
Posterior parts of 3 vertebral columns of which darker specimens (Figure 2D) represent *Gila elegans*.

N.B. All of above remains are non-mineralized, either because of their relative recency or dryness of the cave environment. The bones are very well preserved; some might be at least as old as those of the California condor.

8. Cave passageway — pack rat's nest to North. June 18, 1969, Robert Page.
5 fossil cyprinid vertebrae (articulated), tentatively referred to *Gila*. From caudal part of vertebral column of a large minnow around 400 mm S.L.

9. E-W Tr, Grid GG, 5-10 cm. June 23, 1939, Lawrence Powers, Robert Euler, Robert Page.
Gila elegans (probably)
Complete urohyal, 27.5 mm T.L. From a fish ca. 330 mm S.L.

References

Behnke, R.J., and D.E. Benson
1980 Endangered and threatened fishes of the upper Colorado River basin. Coop. Ext. Serv., Colorado State University, Bulletin 503A:1-34, illus.

Gehlbach, F.R., and R.R. Miller
1961 Fishes from archaeological sites in northern New Mexico. Southwestern Nat., 6:2-8, figs. 1-2.

Holden, P.B., and W.L. Minckley
1980 *Gila robusta* Baird and Girard. Roundtail chub, p. 172. *In:* Lee, D.S., et al., Atlas of North American Freshwater Fishes. Raleigh: N. Caro. State Mus. Nat. Hist.

Holden, P.B., and C.B. Stalnaker
1970 Systematic studies of the cyprinid genus *Gila*, in the upper Colorado basin. *Copeia*, 1970(3): 409-420, figs. 1-4.

Miller, R.R.
1946 *Gila cypha*, a remarkable new species of cyprinid fish from the Colorado River in Grand Canyon, Arizona. Jour. Wash. Acad. Sc., 36(12):409-415, fig. 1.
1955 Fish remains from archaeological sites in the lower Colorado River basin, Arizona. Pap. Mich. Acad. Sci., Arts, and Letters, 40(1954): 125-136, fig. 1, pls. 1-5.

Minckley, W.L., and N.T. Alger
1968 Fish remains from an archeological site along the Verde River, Yavapai County, Arizona. Plateau, 40(3):91-97.

Minckley, W.L., and J.E. Deacon
1968 Southwestern fishes and the enigma of "Endangered Species." Sci., 159(3822):1424-1432, figs. 1-3.

Smith, G.R., R.R. Miller, and W.D. Sable
1979 Species relationships among fishes of the genus *Gila* in the upper Colorado River drainage, pp. 613-623. *In* Linn, R.M., ed., Proc. First Conf. Res. Nat. Parks. Vol. 1, U.S. Nat. Park Serv. Trans. & Proc. Ser. No. 5.

Chapter 6

Ungulate Remains from Stanton's Cave: An Identification List

by

C.R. Harington
Curator of Quaternary Zoology
National Museums of Canada, Ottawa

Introduction

First, a brief explanation of how I became involved in this project. The fossil history of the mountain goat *(Oreamnos)* is very poorly known and I was fortunate enough to find and identify an intriguing undescribed cranial fragment belonging to that genus in the fossil vertebrate collection of the National Museums of Canada. It had been collected about 1932 from what are considered to be deposits of the last (Sangamon) interglacial near Quesnel Forks, British Columbia. Evidently this is the oldest known specimen of *Oreamnos* (Harington 1971). This find led me to review previous reports of Pleistocene mountain goats, including the relatively small extinct species *Oreamnos harringtoni* from southwestern North America. Because of my interest in that species, Paul Martin asked me to accompany him on an expedition to Rampart Cave, Arizona, in April 1971 in order to help collect and identify mammalian bones from his new excavation, which was carefully controlled stratigraphically and geochronologically. Many *Oreamnos harringtoni* bones, and some bones of other mammals, were recovered and briefly described (Harington 1972). Soon after, at Bob Euler's request, I agreed to examine and identify a collection of ungulate remains from Stanton's Cave, Arizona, that arrived in Ottawa in October 1972. The following list of identifications, with comments on the preservation of the elements, measurements, and the ages of individuals (where this could be determined), was completed in February 1973.

Some of the mountain goat material seems rather large for typical *Oreamnos harringtoni*, and I have generally designated it as *Oreamnos* sp., with a remark relative to the living species *Oreamnos americanus*. This is interesting because the Rampart Cave material I have seen in relatively uniform in its small size, and is more easily identified as *Oreamnos harringtoni*. I should mention also that many specimens have been gnawed by small rodents (possibly pack rats?), some fit together, some are charred; and in one case (specimen 39) soft tissues are preserved.

Of 120 specimens identified, approximately 80% are best referred to mountain goat (about 58% being referable to Harrington's Mountain Goat *[Oreamnos harringtoni])*, 13% to Bighorn Sheep *(Ovis canadensis)*, 5.5% to horse *(Equus* sp. — it is worth noting that one of these specimens represents a small asslike horse, possibly *Equus [Asinus] conversidens)*, and 0.5% to bison *(Bison* sp. — cattle *[Bos taurus]* must be considered, depending upon the geological age of the specimens, but Euler [personal communication 1973] considers "the chances of *Bos* being in the cave are remote . . ."). It is worth noting that many of the mountain goats whose remains are preserved in the cave evidently died before reaching maturity.

When did Harrington's Mountain Goat occupy Stanton's Cave? Eleven radiocarbon dates on horn sheaths of that species indicate that it lived in the region from at least 20,060 ± 540 years B.P. (A-3428) to about 11,700 years B.P. (A-3437): the mean of the ten finite dates being about 14,500 years B.P.

This collection of ungulate remains from Stanton's Cave, along with a partial mountain goat cranium *(Oreamnos* sp., probably *O. harringtoni)* from Prayerstick Cave — located between Rampart Cave and Stanton's Cave — was returned to Jim Mead at the Department of Geosciences, University of Arizona, Tucson, in December 1981. I have not commented further upon the implications of the mountain goat material because of Mead's (1983) excellent and comprehensive study of *Oreamnos harringtoni.*

Ungulate Remains: a List of Identifications with Notes

Locality — pack rat's nest near FS #3 (6/16/69)

1. *Oreamnos harringtoni* — neural spine fragment of anterior thoracic vertebra — ?immature.
2. *Oreamnos harringtoni* — second phalanx (medial) of left ?hind foot — ?immature.
3. *Oreamnos harringtoni* — second phalanx (lateral) of left ?hind foot — immature.
4. *Oreamnos harringtoni* — third phalanx or hoof (medial) of left ?hind foot — ?immature.
5. *Oreamnos harringtoni* — limb bone fragments (A,B).

Locality — Grid BB, 0-5 cm (7/2/69)

6. ?*O. harringtoni* — proximal end of anterior (?third) right rib, ?immature.

Locality — area on talus outside of fence (6/14/69)

7. *O. harringtoni* — left metatarsal (distal epiphysis missing). Probably less than 2 years old.

Locality — surface — antechamber #1 — area to N. of passageway from main room to next room back — found in rocks (6/19/69).

8. *O. harringtoni* — distal anterior portion of right humerus (coronoid fossa and part of shaft above). Heavily worn.
9. Unidentified.

Locality — main room N-S Test Trench, Grid K-19, 5-30 cm

10. Unidentified — ?fragment of a second phalanx of a mammal larger than a mountain goat.
11. *O. harringtoni* — anterior face (near midshaft) of left tibia.
12. *O. harringtoni* — second phalanx (medial) of left ?forefoot. Immature (proximal epiphysis lacking).
13. *O. harringtoni* — worn fragment of left (lateral) third phalanx or hoof.
14. *O. harringtoni* — coronoid process fragment of right ramus of mandible.
15. *O. harringtoni* - LP^3. From an old animal (well worn).
16. Unidentified.
17. *O. harringtoni* — proximal sesamoid (lateral) left forelimb.
18. *O. harringtoni* — distal sesamoid (lateral) left forelimb.

Locality — main room N-S Test Trench, Grid H-19 (6/23/69)

19. ?*Ovis canadensis* (or about that size of mammal) — lateral face of a second phalanx. Heavily eroded.

Locality — Grid HH, 0-5 cm (6/23/69)

20. Unidentified (charred bone?).
21. Tentatively referred to *O. harringtoni* — left radio-ulna (distolateral fragment).

Locality — main room N-S Test Trench, Grid K-19, 0-5 cm (6/27/69)

22. *Ovis canadensis* — left metatarsal (lateral condyle).
23. *Equus* sp. — appears to be a proximal phalanx of the forefoot representing a relatively small horse such as *Equus (Asinus conversidens)*.

Measurements (mm)

Total length	73.8+ heavily eroded
Proximal width	35.2+
Proximal depth	25.7+
Midshaft width	21.8+
Midshaft depth	17.8+
Distal width	30.5+
Distal depth	13.7+

Locality — pack rat's nest near FS #3 and FS #5 (6/17/69)

24. *O. harringtoni* — fragment (including much of the centrum, lacking epiphysis) of a lumbar vertebrae. Immature.
25. *O. harringtoni* — fragmentary third phalanx or hoof of an ?immature or small individual.
26. Unidentified + seven bone scraps (unidentified).
27. *Ovis canadensis* — first phalanx of hind limb.

Locality — surface, SE quadrant

28. ?*Ovis canadensis* — second phalanx (medial) of left ?hind limb.

Locality — main room N-S Test Trench, Grid J-19, 5-30 cm (6/27/69)

29. Unidentifiable limb bone fragment of medium-sized mammal.
30. *Bison* size. Semi-lunar bone of carpal.

Locality — base and horn from pack rat nest in upper end of antechamber I (6/17/19)

31. Unidentified — appears to be fragment of (ventral region) of an atlas vertebrae. Try *Ovis canadensis* or deer.
32. *O. harringtoni* — right metacarpal (distal epiphysis present but not fused to shaft). Immature — probably from an individual less than 2 years old.

Measurements

T.L.	91.8 mm	M.W.	21.5 mm
P.W.	30.1 mm	M.D.	12.2 mm
P.D.	19.8 mm	D.W.	34.8 (above condyles) 32.9 (at condyles)
		D.D.	19.0

33. *O. harringtoni* — first medial phalanx of left forelimb.

Measurements

T.L.	37.8 mm	M.W.	16.0 mm
P.W.	18.4 mm	M.D.	13.8 mm
P.D.	19.3 mm	D.W.	18.1 mm
		D.D.	14.5 mm

34. ?*O. harringtoni* — first phalanx of hind limb.

Measurements

T.L.	36.9 mm	M.W.	12.9 mm
P.W.	14.9 mm	M.D.	11.9 mm
P.D.	17.8 mm	D.W.	13.4 mm
		D.D.	12.8 mm

35. *O. harringtoni* — second phalanx of ?forelimb.

Measurements

T.L.	36.9 mm	M.W.	13.0 mm
P.W.	17.0 mm	M.D.	11.9 mm
P.D.	18.2 mm	D.W.	14.1 mm
		D.D.	16.3 mm

36. ?*O. harringtoni* — second phalanx — ?pathological growth near proximal end. Gnawed by small rodent.
37. *O. harringtoni* — fragment of distal end of left metacarpal.
38. *O. harringtoni* — condyle of distal epiphysis of metatarsal — from an immature individual.
39. *O. harringtoni* — distal epiphysis of metatarsal and attached first phalanges. Some soft parts preserved.

Measurements

D.W. 34.0 mm (above condyles)
D.D. 31.0 mm (at condyles)
D.D. 19.8 mm

39b. *O. harringtoni* — first phalanx

Measurements

T.L.	41.9 mm	M.W.	12.7 mm
P.W.	15.4 mm	M.D.	12.7 mm
P.D.	18.4 mm	D.W.	14.2 mm
		D.D.	13.8 mm

39c. *O. harringtoni* — first phalanx (similar dimensions)

40. Unidentified (try large mountain goat or bighorn sheep). First phalanx of hind limb.

Measurements

T.L. 53.8 mm	M.W. 15.3 mm
P.W. 19.5 mm	M.D. 15.0 mm
P.D. 21.8 mm	D.W. 13.0 mm
	D.D. 14.4 mm

41. *O. harringtoni* — right astragalus. Gnawed by rodents. Nearly as large as a recent specimen of *O. americanus* (NMC 13757).

Measurements
T.L. 36.5 mm
M.D.20.1 mm
Max W. 25.0 mm

42. *O. harringtoni* — right radius (distal end missing).

Measurements

T.L. —	M.W. 18.3 mm
P.W. 32.1+ mm	M.D. 11.2 mm
P.D. 17.6+	D.W. —
	D.D. —

43. *O. harringtoni* — fragment of area near top of olecranon fossa of right humerus.
44. *O. harringtoni* — shaft of left femur. From an immature individual.
45. *O. harringtoni* — part of head and neck of left femur. Gnawed by small rodent.
46. *O. harringtoni* — horn sheath (basal area missing). Apparently a young animal or female.
47. ?*O. harringtoni* — fragment (neural spine and area around neural canal) of anterior thoracic vertebra. ?Juvenile.
48. Unidentified. 9 fragments including parts of auditory capsules and posterior cranium.
49. Unidentified — limb bone shaft.

Locality — main room N-S Test Trench, Grid I-19, 15-20 cm (6/24/69)

50. *Oreamnos* sp. — frontal bone fragment at base of left horncore.

Locality — Grid HH, 25-50 cm (6/25/69)

51. *O. harringtoni* — horncore tip fragment.
52. *O. harringtoni* — horncore fragment.
53. *O. harringtoni* — third (medial) phalanx or hoof.

Locality — bone from pack rat nest adjacent to F.S. #13 (6/18/69)

54. Unidentified — patella (sheep or goat size).

Measurements
L. 45.0 mm
W. 29.8 mm
H. 22.8 mm

55. *O. harringtoni* — distal fragment of left humerus.
56. Unidentified — ?limb bone fragment.
57. ?*Ovis canadensis* — first phalanx (medial) of right metatarsal.

Measurements

T.L. 55.2 mm	M.W. 13.3 mm
P.W. 17.7 mm	M.D. 13.6 mm
P.D. 19.5 mm	D.W. 16.1 mm
	D.D. 13.2 mm

58. *O. harringtoni* — third phalanx (lateral) or hoof of hind foot.

Measurements
L. 45.7 mm
W. 13.5 mm
H. 19.2 mm

59. *O. harringtoni* — third phalanx (medial) or hoof of hind foot.

Measurements
L. 47.2 mm
W. 13.3 mm
H. 18.6 mm

(58 and 59 seem to be a pair.)

Locality — main room N-S Test Trench, Grid H-19, 5-10 cm (6/23/69)

60. *O. harringtoni* — right horncore (proximal area missing). Gnawed by small rodent.

Locality — from pack rat nest at FS #22 (6/20/69)

61. *O. harringtoni* — split horncore (proximal area missing).
62. *O. harringtoni* — fragments (on right side) of first and second vertebrae of sacrum.

Locality — pack rat's nest near FS #3 and FS #5 (6/17/69) (bones charred)

63. ?*Ovis canadensis* — first (lateral) phalanx of ?hind foot. Appears to be from an old animal.

Measurements

T.L. 58.3 mm	M.W. 18.7 mm
P.W. 21.2 mm	M.D. 16.8 mm
P.D. 24.0 mm	D.W. 25.7 mm
	D.D. 18.3 mm

64. *O. harringtoni* — eroded, charred centrum of axis vertebra. Plus 8 fragments — probably from the same vertebra.
65. *O. harringtoni* — anterior thoracic vertebra. Charred, broken and eroded.

Locality — surface NW quadrant

66. *Ovis canadensis* — distal half of the left tibia.

Measurements
D.W. 39.5 mm
D.D. 30.9 mm

67. *O. harringtoni* — hornsheath (proximal area missing) and fragment (probably from near base of this specimen).
68. Unidentified — limb bone fragment of medium-sized mammal. Heavily pitted. Plus fragment.
69. ?*Oreamnos* — ?right scaphoid.
70. *O. harringtoni* — eroded right astragalus.

Measurements
L. 32.2 mm
W. 20.6 mm
H. 16.9 mm

71. *Oreamnos* — distal end of left femur.

Measurements
D.W. 44.4 mm
D.D. 55.7mm

72. ?*Equus* sp. — rib fragment. Approximately the size and shape of the right third rib of a horse.
73. *Equus* sp. — third phalanx or hoof of a small to medium-sized horse. Evidently slightly larger than *Equus (Asinus) conversidens*.

Measurements
L. 62 mm (approximately)
W. 70.8 mm
H. 44.8 mm

Locality — cave passageway, pack rat's nest (including FS #12) (6/18/69)

74. *Ovis canadensis* — left metacarpal. Immature — probably slightly less than two years old because the distal epiphysis is not fused to the shaft.

Measurements

T.L.	170.5 mm	M.W.	17.2 mm
P.W.	33.1 mm	M.D.	14.2 mm
P.D.	21.4 mm	D.W.	34.5 mm (above condyles) 35.5 mm (at condyles)
		D.D.	21.5 mm

75. ?*Ovis canadensis* — right metatarsal. Seems to be derived from a juvenile.
76. *Ovis canadensis* — right metatarsal — part of shaft of an immature.
77. *Ovis canadensis* — first phalanx of hind limb. Proximal epiphysis not fused to shaft — probably from an immature individual.

Measurements
T.L. 51.4 mm (approximately)

78. *O. harringtoni* — left humerus (distal half).
79. *Oreamnos* sp. — right calcaneum. Gnawed.
80. *Ovis canadensis* or *Oreamnos* sp. — right humerus (condylar fragment) and piece above (separate).
81. Unidentified — possibly a fragment of 80, but no "fit" is obvious.
82. *O. harringtoni* — right metatarsal. Proximal articulation lacking in part. Distal epiphysis missing. Probably less than 2 years old.

Measurement
L. 33.3 mm

83. *O. harringtoni* — right astragalus.

Measurements
L. 33.6 mm
W. 23.1 mm
D. 19.4 mm

84. ?*Oreamnos* sp. — right astragalus.

Measurements
L. 42.8 mm
W. 28.6 mm
D. 24.0 mm

85. *O. harringtoni* — first (medial) phalanx of forelimb.
86. *O. harringtoni* — first (lateral) phalanx of left forelimb.
87. *O. harringtoni* — first (lateral) phalanx of left forelimb.
88. *O. harringtoni* — second phalanx of forelimb — distal end broken, proximal end lacking epiphysis. Immature.
89. *Oreamnos* sp. — right hornsheath. Seems rather large for *O. harringtoni*. From an individual estimated to be 8 years old. Stained a reddish color.

Locality — from pack rat nest (7/02/69)

90. *Bison* sp. — part of "cuticle" on outer surface of third phalanx or hoof.

Locality — pack rat's nest near FS #3 (6/16/69)

91. *Bison* sp. — anterior facet of acetabulum of right pelvic bone. Deep depression toward shaft of ilium is almost identical to that in a right pelvic fragment of *Bison crassicornus* (NMC 11690), to which it was compared. Generally similar to *Bison* in size and shape.

Locality — Grid H-19, 0-5 cm

92. ?*Oreamnos* or *Ovis* — possibly a fragment of the femur shaft.
93. Unidentified — Possibly a skull fragment of a mountain goat or sheep.

Locality — main room N-S Test Trench, Grid I-19, 20-25 cm

94. Unidentified. Possibly part of the shaft of the right radius of a horse.
95. ?*Oreamnos* sp. — second phalanx of hind limb.

Measurements
T.L. 34.8 mm

Locality — from pack rat nest adjacent to FS #13 in upper end of antechamber #1. Nest was hard packed on top and horns came from level about 10-15 cm (6/18/69).

96. Unidentified — cuticular material. Part of a male mountain sheep hornsheath near tip?
97. *Oreamnos* sp. — part of a hornsheath, which seems unusually robust for *O. harringtoni* and which seems more like *O. americanus* — from an adult male.
98. *O. harringtoni* — nearly complete hornsheath.
99. *O. harringtoni* — part of hornsheath with unusual spatulation at tip.
100. *O. harringtoni* — part of hornsheath near tip.
101. *O. harringtoni* — part of hornsheath near tip.
102. *O. harringtoni* — part of hornsheath near tip, with most of horncore inside.

Locality — material found along large log in southwest quadrant adjacent to FS #34, 35. (7/5/69)

103. *Ovis canadensis* — right hornsheath near tip. Proximal area of fragment measures approximately 45.2 x 33.6 mm (soil sample taken from this sheath for P. Martin). Probably from a male.
104. Unidentified — series of 3 fragmentary (and centrum of a fourth) articulated cervical vertebrae.

Locality — bone from pack rat's nest adjacent to FS #14 in upper end of antechamber #1.

105. *O. harringtoni* — left metatarsal of an adult.

Measurements

T.L.	98.3 mm	M.W.	16.3 mm
P.W.	23.6 mm	M.D.	12.3 mm
P.D.	19.2 mm	D.W.	27.7 mm
		D.D.	17.9 mm

106. *Oreamnos* sp. — proximal end of shaft of tibia (proximal epiphysis missing). Probably no more than 3-½ years old.
107. ?*Oreamnos* sp. — part of proximal end of right metacarpal.
108. Unidentified. Possibly a fragment of mountain goat or sheep bone near distal end of humerus or femur.
109. *O. harringtoni* — right astragalus.

Measurements

L.	33.3 mm
W.	22.4 mm
H.	19.1 mm

110. *O. harringtoni* — right naviculo — cuboid.
111. *O. harringtoni* — right cuneiforms (lateral and middle cuneiforms fused) and medial cuneiform.

(109, 110, and 111 articulate and are probably from the same individual.)

112. ?*Oreamnos* sp. — second phalanx of hind foot.
113. Unidentified — anterior lumbar vertebra (lacking epiphyseal plates on centrum). Immature individual.
114. Unidentified — anterior lumbar vertebra (lacking epiphyseal plates on centrum). Immature individual.

(113 and 114 probably from same individual.)

115. Unidentified — fragment possibly from the braincase of a mountain goat or sheep.
116. *O. harringtoni* — right pelvic fragment including acetabulum and parts of the ilium, ischium and pubis.

Measurements

L. of acetabulum	27.7 mm
W. of acetabulum	26.8 mm
W. of "neck" of ilium	16.9 mm
W. of "neck" of ischium	20.1 mm
W. of "neck" of pubis	10.8 mm

117. Unidentified — distal part of shaft (posterior) of right humerus. About mountain goat or sheep size.
118. *Oreamnos* sp. — fragment of right calcaneum. Tuber calcis lacking.
119. Unidentified — left radius (eroded). Seems to be longer than *Oreamnos*, possibly referable to *Ovis canadensis*.
120. Unidentified — probably shaft fragment of right radius (note ulnar facet) of a mountain sheep or goat. Charred.
121. ?*Oreamnos harringtoni* - right tibial shaft of a juvenile or immature individual.

Locality — miscellaneous bone — surface of antechamber #1 (6/17/69)

122. *Equus* sp. — heavily eroded centrum and bases of transverse processes of ?fifth cervical vertebra.
123. *Equus* sp. — heavily eroded centrum and bases of transverse processes of fifth lumbar vertebra.
124. Unidentified — possibly a flattened portion of a hornsheath of a mountain sheep.
125. Unidentified — cuticular sheath of an ungulate hoof. Possible *Bison* considering the thickness of the sheath and steepness of the "rise" on the outer surface.
126. *Oreamnos* sp. — parts of frontal bone (including edge of orbit) and lacrimal bone.
127. Probably *Equus* sp. — a central thoracic (approximately fifth to ninth) vertebral fragment.
128. *O. harringtoni* — hornsheath fragment.
129. *Oreamnos* sp. — proximal end of right tibia. Adult.
130. ?*Oreamnos* sp. — distal epiphysis of left radio-ulna. Probably from an animal less than 3 years old.
131. Unidentified — possibly distal part of femoral shaft of a mountain sheep. Immature.
132. *Ovis canadensis* — proximal half of right metatarsal.
133. *Oreamnos harringtoni* — anterior half of left ramus of mandible (includes diastema, and LP_2-LM_2 Mature.

Measurements

Alveolar length		
	P_2-P_4	24.7
	M_1-M_2	32.0
	P_2-M_2	57.0

	L	W
P_2	6.9	5.5
P_3	8.1	6.4
P_4	10.2	7.9
M_1	13.7	8.9
M_2	17.8	9.6

134. *O. harringtoni* — left ulna (olecranon broken off).

Measurements

Maximum width	21.3 mm
Depth of "neck" of shaft	8.2 mm

135. *O. harringtoni* — lower half of right tibia.
136. *O. harringtoni* — two articulating anterior thoracic vertebrae. Fragmentary with all but one of the epiphyseal plates lacking.
137. Unidentified — possibly a fragment of ?cranial bone near auditory capsule. Plus about 11 small unidentified fragments.

Locality — bone found in back part of cave in pack rat's nest (6/16/69)

138. *Oreamnos* sp. — left metacarpal (lacking distal epiphysis). About the size of *Oreamnos americanus*.

Measurements

P.W. 32.0 mm M.W. 23.5 mm
P.D. 21.8 mm (approximate)
M.D. 13.6 mm

139. *Oreamnos* sp. — fragment of ?third lumbar vertebra. Epiphyseal plates are lacking. Plus two small unidentified fragments.
140. Fragment of a ?snake skeleton.

Locality — main room N-S Test Trench, Grid H-19, 15-20 cm (6/24/69)

141. ?*Oreamnos* sp. heavily eroded fragment of a left astralagus.

Locality — southwest quadrant — surface.

142. *Ovis canadensis* — distal end (condylar region) of a right femur. Heavily eroded.

Locality — main room N-S Test Trench, Grid G-19, 5-10 cm (6/27/69)

143. Unidentified — appears to be a limb bone (right femoral shaft?) fragment of medium-sized mammal.

Locality — pack rat's nest near FS #3 and #5 (6/17/69)

144. *Equus* sp. — ball joint of left femur.
145. ?*Oreamnos* sp. — fragmentary right naviculocuboid.

Locality — pack rat's nest near FS #3 and FS #5 (6/17/69)

146. ?*Oreamnos harringtoni* — heavily worn thoracic vertebra fragment.
147. Unidentified — probably part of a limb bone.

Locality — antechamber #1 surface (7/1/69)

148. *O. harringtoni* — right tibia (proximal epiphysis not fully fused to shaft and distal epiphysis missing). Probably from an individual about 3 years old.
149. *Ovis canadensis* — left astragalus (gnawed).
150. *Ovis canadensis* — left calcaneum (most of tuber calcis lacking).

(149 and 150 articulate and probably from the same individual.)

151. Unidentified — bone fragment.

Locality — bone from pack rat's nest adjacent to FS #14 (6/18/69)

152. ?*Oreamnos harringtoni* — fragmentary left ramus of mandible of a juvenile. Probably less than 2 years old. The milk teeth dP_2-dP_4 have fallen out of the sockets. M_1 was probably in wear, while the remaining molars had not erupted. M_1 appears to be larger than teeth at a similar stage in a comparative specimen of *Oreamnos americanus* (NMC 16028).

Locality — north of passageway from main room to next room back, found in rocks (6/19/69).

153. *Oreamnos harringtoni* — damaged left scapula.

Measurements

T.L. of glenoid-tuber scapulae	38.9 mm
Min. A.P. diam. at "neck"	22.4 mm
W. of glenoid cavity	27.9 mm

154. Unidentified — bone fragment.

Acknowledgments

I am grateful to Dr. Paul S. Martin (Department of Geosciences, University of Arizona) for sparking my interest in this topic, and to Dr. Robert C. Euler (National Park Service) for allowing me to describe the ungulate remains from Stanton's Cave. Austin Long supplied the radiocarbon dates on *O. harringtoni*. I thank Dr. C.G. van Zyll de Jong (Curator of Mammals, National Museums of Canada) for allowing me to borrow several comparative specimens.

Table 6-1. Radiocarbon dates of horn sheaths of Harrington's Mountain Goat *(Oreamnos harringtoni)* from Stanton's Cave.

Date (B.P.)	Laboratory Number
11,700 ± 300 (380)*	A-3437
11,920 ± 760 (810)*	A-3439
12,090 ± 550	A-3435
12,310 ± 420	A-3427
13,480 ± (500)*	A-3436
14,100 ± 1000	A-3431
›14,500	A-3429
15,700 ± 1400 (1800)*	A-3438
16,090 ± 380	A-3430
17,600 ± 1500	A-3432
20,060 ± (930)*	A-3428

These dates courtesy of the National Science Foundation Accelerator Facility for Radioisotope Analysis at the University of Arizona, Tucson.

*The best standard deviation.

References

Harington, C.R.

1971 A Pleistocene mountain goat from British Columbia and comments on the dispersal history of *Oreamnos*. *Canadian Journal of Earth Sciences* 8(9):1080-1093.

1972 Extinct animals of Rampart Cave. *Canadian Geographical Journal* 85(5):178-183.

Mead, Jim I.

1983 Harrington's extinct mountain goat *(Oreamnos harringtoni)* and its environment in the Grand Canyon, Arizona. Ph.D. Dissertation. Tucson: University of Arizona.

Chapter 7

The Bird Bones from Stanton's Cave

by

Amadeo M. Rea

Curator of Birds and Mammals
San Diego Natural History Museum

and

Lyndon L. Hargrave (deceased)

Approximately one thousand bird bones were excavated from Stanton's Cave in the Grand Canyon, about 51 river kilometers SSW of Lees Ferry, Coconino County, Arizona. Approximately 600 bones are sufficiently complete to permit species identification. Although large Pleistocene-Holocene bone deposits containing numerous avian remains have been recovered from a number of dry caves in southern New Mexico, Stanton's Cave represents the largest avian fossil collection obtained to date from an Arizona site. The Stanton's Cave material is not mineralized, but is excellently preserved except for occasional breakage by carnivores, trampling by herbivores, or gnawing by pack rats *(Neotoma* sp.).

Most of the larger bones were identified by Hargrave, while the smaller bones were identified by Rea. The authors together re-examined most of the unusual and extralimital species. Identifications of teratorns and condors were made by Hildegarde Howard, Department of Vertebrate Paleontology, Los Angeles County Museum, California. We are grateful to Dr. Howard for permitting us to incorporate her findings into this report.

Due to the careful excavation methods employed in Stanton's Cave, a considerable number of very small bird bones were recovered. Most of these belong to the Order Passeriformes (perching or song birds), the most taxonomically diverse avian order. Relatively little is known of the fossil history of the passerines, owing to their usually small size, infrequency of preservation, as well as the difficulty in identification because of the numbers of closely related species and their poor representation (as well as poor preparation) in modern osteological collections. During historic times many passerine birds have been dynamically extending or contracting their ranges. This is particularly noticeable along the United States-Mexico International Boundary (see, for instance, Phillips 1968; Rea 1983). Also, small birds (as well as amphibians and reptiles) often have highly restricted ecological requirements, sometimes even during migrations, hence they are useful tools in paleoecological and paleoclimatic reconstructions.

Two sections of Stanton's Cave were selected for detailed (fine screen) stratigraphic excavation. One trench sampled an area of 8 square surface meters, the other trench almost 12 square meters. On the basis of bighorn and extinct mountain goat dung analysis, Paul S. Martin (personal communication) considers the top of the 20-25 cm stratum to represent the end of the last glacial period, with all materials above 20 cm representing the post-glacial accumulation. Pollen profiles confirm this demarcation, but plant cuticles from the dung at this point appear not to reflect a vegetational change. (Selective browsing by bighorn sheep may obscure actual vegetational changes that occurred in response to climatic change.)

The only reasonably certain Recent material from the cave is from the 0-20 cm strata. "Surface" materials cannot be considered exclusively modern in origin, since large-sized condors *(Gymnogyps amplus),* typical of the Pleistocene appear in the surface accumulation. Holocene pack rat middens likely contain some Recent bones along with scavenged Pleistocene bones, but it is not possible to segregate these without vegetational or radiocarbon analyses of each midden. Considering the 38-thousand-plus-year history of the accumulation, it is likely that the majority of the pack rat material is of Pleistocene origin.

Except for the enigmatic split-twig figurines, dating between 3000 and 4000 B.P., and a few miscellaneous Anasazi artifacts, there is no evidence of human agency in bringing biological materials into the cave.

The taxonomy and sequence of families follows Rea (1983) and various sources cited therein. Certain well-known names that have been changed for taxonomic or nomenclatural reasons are given in parentheses.

Numbers in parentheses refer to the specimen field number. All bones are now catalogued in the paleontological collection of the Museum of Northern Arizona (MNA), Flagstaff.

The following abbreviations are used in citing modern comparative specimens:

AMR	Amadeo M. Rea Collection at San Diego, California
ARIZ	University of Arizona Collection, Tucson
LLH	Lyndon L. Hargrave Collection at MNA
MVZ	Museum of Vertebrate Zoology, Berkeley
SD	San Diego Natural History Museum, San Diego

Species Accounts

Podicipedidae

Eared Grebe, *Podiceps nigricollis.* A single femur (SC 246) was obtained from the surface deposit. The species is a common transient in the Southwest.

Western Grebe, *Aechmorphorus occidentalis.* Two elements were obtained from pack rat middens. Their fragmentary nature precluded measurement and subspecific determination. This species is a common transient and winter resident.

Teratornithidae

Merriam's Teratorn, *Teratornis merriami.* Two specimens of this enormous Teratorn were obtained: a partial cranium (SC 69) and a complete right humerus (SC 144); Dr. Howard made the identifications. *T. merriami* has been recorded from San Josecito Cave, Nuevo León, México, in addition to the Rancho La Brea, McKittrick, and Carpinteria asphalt pits of California. Although the Pleistocene avifauna of New Mexico is relatively well known, only a *Teratornis* palatine has been discovered there (Arthur Harris, personal communication), and this is the first record for Arizona. An even larger Pleistocene teratorn, *T. incredibilis,* is known from Smith Creek Cave, Nevada (Howard 1952), and Vallecito-Fish Creek area, Anza-Borrego Desert of southern California (Howard 1963, 1972).

The cranium was recovered in the main room (Grid

I-19) in the 25-50 cm stratum, which is late Pleistocene in age. P.S. Martin removed matrix from this bone for pollen analysis. James E. King reported 33.5% *Artemisia,* 16.5% *Juniperus,* and 10.5% *Pinus* pollen, indicating conditions different from those found today in the Canyon (King and Sigleo 1973). The predominance of sagebrush was noted at Rampart Cave (west end of the Grand Canyon) at the 12,000 B.P. level, indicating cooler and moister conditions during the late Pleistocene (Martin et al. 1961).

Measurements (in mm) of the humerus are:

greatest length	308.4
greatest breadth of distal end	58.3
least breadth of shaft	23.6
proximal breadth taken from proximal edge of bicipital contour	62.1
proximal breadth across external and internal tuberosities	58.1
depth of head	23.0
height of ectepicondylar process from distal end	29.3

(All measurements by Howard taken on original bone after restoration.)

After casts were made at Los Angeles County Museum, the humerus was submitted to Austin Long, Department of Geosciences, University of Arizona, for radiocarbon dating. The age of 15,230 ± 240 B.P. (A-1238) is the first direct age determination on this extinct species (Euler 1978:159). Dr. Long notes the high organic fraction was extracted with acetone to remove glyptal cement and then water to remove possible pack rat urine. P.S. Martin (personal communication to Rea), who excavated this humerus, points out that conditions of deposition were especially favorable for obtaining a reliable carbon age. The deposit was enclosed in a pack rat midden sealed from above, with no evidence of water percolation. The bone had a very fresh appearance and associated mammal dung was dusty and excellently preserved. Chance contamination after placement in the midden was minimal.

The two *Teratornis* bones might be from a single individual. The radiocarbon date of the humerus, the stratigraphic level of the cranium, and the pollen matrix all correspond.

Vulturidae (Cathartidae *auctorum)*

Turkey Vulture, *Cathartes aura.* At least four individuals were recovered, one (referred) a juvenal. *Cathartes* appears to become increasingly common at the end of the Pleistocene with the demise of such large scavenging birds as *Gymnogyps* and *Teratornis* (see Howard 1962; L. Miller 1960b). *Coragyps,* the other genus of small New World vulture, is unrepresented in Stanton's Cave, though its Pleistocene "species," *C. occidentalis* (probably only a temporal subspecies), occurs commonly in most other southwestern caves (Brodkorb 1964). R.L. McKenzie and Rea (unpublished data) have recorded this species from the vicinity of Rampart Cave, at the extreme western end of the Grand Canyon.

The Turkey Vulture breeding today in most of the Southwest is the small *Cathartes a. aura* (predominantly of México and southward), while the more northern *C. a. meridionalis* occurs as a common transient and perhaps occasional winter visitant (Rea 1983:128-130). The subspecific differences are determined primarily by wing lengths. The only measurable wing element from Stanton's Cave is a carpometacarpus (SC 73) taken from a pack rat nest. An axial length of 77.6 mm indicates a very small representative of the southern subspecies. Sixteen breeding season *C. a. aura* from Arizona and New Mexico localities in the AMR Collection measure 76.1-83.9, $\bar{X}$ = 81.11 mm. This suggests that the smaller Turkey Vulture may have bred north of the international border even in Pleistocene times, as was suspected by Wetmore (1935). Wing elements from known-age deposits are needed to examine this hypothesis. Hargrave (1970) has identified feathers of the small race from a Basketmaker site in southern Utah. Rea identified a humerus (149.4 mm) of the large northern *C. a. meriodionalis* from Vulture Cave at the west end of Grand Canyon (Mead and Phillips 1981). It was presumably a migrant.

California Condor, *Gymnogyps* sp. Condors are represented by at least 68 individual bones in addition to those reported earlier by deSaussure (1956) and Parmalee (1969). Both *G. amplus* and *G. californianus* are represented. The "ancestral" California Condor, *G. amplus,* averages larger than the modern condor and shows slight qualitative differences in the skull (see Fisher 1944, 1947). Whether *G. amplus* is a distinct species or a chronocline of the living species is debated. There is a great deal of overlap in postcranial measurements. Howard notes that diagnostic measurements could be made on 30 condor bones from Stanton's Cave. Seventeen of these are larger than the maximum for the present-day population, and fall within the range of *G. amplus.* Thirteen fall within the overlapping ranges of the two taxa. The only condor bone that is clearly the modern taxon (i.e., too small for *G. amplus* and outside the range of overlap) is the right coracoid from the surface reported by Parmalee (1969) as ca. 104 mm. The minimum *G. amplus* coracoid is 111 mm (Fisher 1947).

Gymnogyps amplus is not know to be temporally sympatric with the modern condor and we favor considering it a subspecies that was the progenitor of *G. californianus.* Table 7-1 shows, however, some apparent stratigraphic overlap between the two taxa in Stanton's Cave. P.S. Martin (personal communication) believes that the two *G. amplus* bones obtained on the surface deposit in this excavation and the third surface tibiotarsus reported by Parmalee (1969) measuring 228 mm, must not be in primary association. Assuming that strata above the 20 cm level are post-glacial or Recent, there are only two other problematic specimens of *G.*

amplus. One of these was recorded by the excavators as "5-30 cm" and the other indicated as "15-40 cm," so both could be from Pleistocene strata.

On the basis of size and rights and lefts of separate elements, Dr. Howard notes that at least five individual condors are represented from the cave. Based on bones with stratigraphic proveniences (see Table 7-1), perhaps as many as eight birds are represented.

There is no evidence of juvenal or immature individuals from Stanton's Cave. The species did breed in the Southwest (Wetmore and Friedmann 1933, western Texas; Howard 1962, western New Mexico) and it is known from one immature bone (Miller 1960a) from Rampart Cave at the western end of the Grand Canyon. Perhaps the accessibility of this cave to extinct goats, *Oreamnos harringtoni,* and bighorn sheep, *Ovis canadensis,* made it unsuitable for breeding by these enormous birds. This species requires nearly a year to incubate and fledge its single young, and apparently does not tolerate nest disturbance (Koford 1966).

Ardeidae

Great Blue Heron, *Ardea herodias.* Three bones[1] from pack rat nests are from mature individuals. Three races have been recorded in Arizona (Phillips et al. 1964) but differences involve plumage color rather than size. The largest of the North American herons, it requires open water for fishing and is found commonly in small numbers in the Canyon today.

Accipitridae

Sharp-shinned Hawk, *Accipiter striatus.* One bone, a tibiotarsus in the size range of a female, represents this species which is uncommon in the area today.

Red-tailed Hawk, *Buteo jamaicensis.* The most common true hawk in Arizona today, this species is represented by a complete tarsometatarsus. One burned and broken buteonine pelvis also seems best referred to this species. A coracoid (SC 94) is in the range of overlap between at least four buteonine species, and we do not wish to risk specific identification.

Lesser (common) Black Hawk, *Buteogallus anthracinus.* A distal end of a humerus (SC 85) is clearly this species, which can be distinguished from other buteonine hawks of the Southwest by the shallower olecranal fossa and the more pronounced ectepicondylar prominence.

The Lesser Black Hawk is a summer breeding bird today in Arizona, extending as far as southern Utah (Behle and Perry 1975). It requires flowing water and broad-leaf riparian for nesting. Essentially tropical to subtropical, the species is here at its northern limits. Brodkorb (1964) reports no fossil record. The humerus and a referred phalanx were obtained in pack rat nests, so their age is unknown.

Golden Eagle, *Aquila crysaetos.* Four bones of this species were recovered. A complete but weathered tarsometatarsus obtained on the talus inside but near the mouth of the cave is in the size range of modern adult males. A proximal end of a radius (SC 329) and partial tibiotarsus (SC 119) are immature. Both were obtained on the surface and may represent one individual. A fragment of the distal end of a humerus is from an adult. The lack of eagle bones in the stratigraphic levels suggest that the species was not nesting in the cave, but they could have been nesting on the cliff above the cave and their bones brought in by the pack rats.

Cf. Northern Harrier (Marsh Hawk), *Circus cyaneus.* A partial pelvis (SC 95) from a pack rat nest is tentatively referred to this species. The poor representation of this species suggests a lack of open, marshy or heavily grassed habitat in the area of Stanton's Cave.

Falconidae

Prairie Falcon, *Falco mexicanus.* Four bones (SC 98, 99, 122, 159) of this species represent at least three individuals: an adult, a male-sized immature, and a female-sized immature. We have both noted that of the larger falcons, this species, which nests on cliffs, is most frequently encountered in southwestern archaeological sites. Phillips et al. (1964) consider the Prairie Falcon a statewide resident that was formerly more common.

American Kestrel, *Falco sparverius.* This common small falcon (the "Sparrow Hawk") is represented by three bones, each from a different individual. A humerus (SC 138) is from a nestling. A carpometacarpus (intercarpal space 18.9 mm) is from a female assigned to the nominate race. It is slightly larger than females breeding at higher elevations in Arizona today (AMR Collection). The 25-45 cm level in which it occurred is Pleistocene. Pleistocene breeding populations may have been larger or the element could be from a winter migrant from the north or east. The movement northward of small *F. s. peninsularis* may be a postglacial phenomenon (Rea 1983). A proximal end of a carpometacarpus from the 0-5 cm level appears to be from a small bird, but axial measurements are not possible. A referred ulna from the surface collection (SC 180) appears to be in the size range of male kestrels breeding at higher elevations in the state today.

Tetraonidae

Sage Grouse, *Centrocercus urophasianus.* A left coracoid (SC 133) and left scapula (SC 213) (recovered in association) from the surface of Antechamber One are referred to the same individual. Both are in the female size range, as is a surface humerus reported by Parmalee (1969). A well-preserved right femur (SC 260) is from a pack rat nest in the upper end of Antechamber One. Its minimum axial length is 68.5 mm, maximum 73.4 + mm, putting it also in the size range of a female Sage Grouse. All four bones may be from the same individual.

Pleistocene records of the Sage Grouse in the Southwest are extensive and include Smith Creek Cave, Nevada, and at least five caves in New Mexico (see Brodkorb 1964). The only modern report of the species in Arizona is an observation by Arnold (Huey 1939) in the Mt. Trumbull region, northern Mohave County. The species has suffered considerable range reduction in post-Pleistocene times. Johnsgard (1973) notes that

> at one time this species was found virtually wherever sagebrush *(Artemisia,* especially *A. tridentata)* occurred, throughout many of the western and intermountain states. In early historic times, it occurred in fourteen or fifteen [western] states and was the principal upland game species in nine. However, overgrazing and drought contributed greatly to the species' near demise . . . it is difficult to be optimistic about the long-term future of the sage grouse in North America. The continued clearing of extensive areas of sage for irrigated farming, as has occurred widely in central Washington, and the expanded use of herbicides to improve grazing conditions are likely to further reduce Sage Grouse habitat and populations in further years.

Phasianidae

Extinct Turkey, *Meleagris crassipes.* A distal end of a tarsometatarsus was obtained from a pack rat nest. The nest had been burned by vandals so the bone is somewhat calcined; the fragile specimen was damaged after identification and subsequently restored. It was compared directly with the type series of *M. crassipes* from San Josecito Cave, Nuevo León. The plane of the trochlea and curvature of the distalmost end of the shaft distinguish this paleospecies from the Common Turkey, *M. gallopavo,* occurring in northern Arizona in historic times (see Rea 1980, for a discussion of the interrelationships of these two species). In historic times *M. gallopavo* occurred naturally only as far north as the San Francisco and Lukachukai mountains (Phillips 1946); but it has been successfully introduced on the Kaibab Plateau (Phillips et al. 1964) and in southern Utah (Behle and Perry 1975).

Rallidae

Common Gallinule, *Gallinula chloropus.* The species is represented by the distal end of a tibiotarsus (SC 186). At least in the Southwest, this species does not require extensive vegetation for breeding, only permanent water and aquatic plants. We have not observed the gallinule in the Grand Canyon where there is only one sight record (Monson and Phillips 1981).

American Coot, *Fulica americana.* At least five individuals, represented by 11 bones, were recovered in surface debris or in pack rat nests. The requirements of this species are similar to those of many ducks encountered in the cave deposit.

Scolopacidae

Common Snipe, *Capella gallinago.* One ulna, the size of an adult male, came from the 15-40 cm level of Grid DD. The species is a common transient in Arizona (Phillips et al. 1964).

Long-billed Curlew, *Numenius americanus.* The species, represented by most of the elements of the left wing, was recovered from a pack rat nest. The elements are in the size range of a male. This large shorebird usually occurs in numbers as a transient in open situations. It breeds south to central Utah today, but Phillips et al. (1964) believed that it may have bred formerly as far south as the Grand Canyon.

Spotted Sandpiper, *Actitis macularia.* A right tarsometatarsus, missing the proximal end, was recovered from the 20-25 cm level. It is from an immature individual, based on surface striations. Spotted Sandpipers breed today in the Canyon at least in limited numbers (Carothers and Johnson 1975) and are wide-spread migrants in the states. Bones of the small sandpipers, collectively known as "peeps," were compared with skeletons of *Micropalma himantopus, Tringa solitaria, Actitis macularia, Ereunetes mauri* and *E. pusillis, Erolia melanotos, E. alpina, E. bairdii, E. minutilla, Arenaria melanocephala, A. interpes, Aphriza virgata, and Crocethia alba.* Following Campbell (1979:114), we have conserved the genera *Erolia* and *Ereunetes* (submerged in *Calidris* by the American Ornithologists' Union 1976), pending a detailed osteological study. Though many of these species are superficially similar in the hand, their osteological characters are surprisingly distinctive.

Pectoral Sandpiper, *Erolia (Calidris) melanotos.* The proximal end of right tibiotarsus (SC 363) was excavated from the 30-35 cm level. The tibiotarsus of this species is readily distinguishable from those of other small scolopacid species by: 1) greater slope of internal articular surface and 2) excavation of shaft immediately beneath this. The Pectoral Sandpiper is an uncommon fall migrant chiefly in the lower Colorado River valley, with apparently no other record from northwestern Arizona (Phillips et al. 1964). It is partial to marshy shores (Peterson 1961).

Phalaropodidae

Red Phalarope, *Phalaropus fulicarius.* The proximal end of a left ulna (SC 364) was recovered from the 30-35 cm level. The size and its distinctive characters distinguish this species from both *Lobipes lobatus* and *Steganopus tricolor.* The phalaropes are surprisingly close osteologically to the small sandpipers. This is the rarest of the three phalaropes in Arizona (Phillips et al. 1964).

Northern Phalarope, *Lobipes lobatus.* Complete left and right humeri with referred digit 2, phalanx I, fragment of pelvis, and a fragment of a sternum, were recovered from the 55-60 cm level; a referred left humerus (somewhat damaged) came from the 65-70 cm level. The Northern Phalarope is a common fall migrant in the state (Phillips et al. 1964).

Recurvirostridae

American Avocet, *Recurvirostra americana.* A right coracoid was recovered from a pack rat midden. Avocets are common migrants in the state (Phillips et al. 1964) and there have been recent nesting attempts on the lower Salt River south of Grand Canyon in central Arizona (Rea 1983).

Laridae

Franklin's Gull, *Larus pipixcan*. The distal end of a humerus (SC 121) and a partial ulna (SC 132) were recovered from the cave in nonstratigraphic contexts (on rocks and in a pack rat midden, respectively). The fossils are smaller than *L. delawarensis*, the more common species today, and they differ in characters from *L. philadelphia, L. atricilla, L. heermanni, Xema sabini*, and *Rissa tridactyla*, the other small to moderately small sized gulls to be expected in the interior Southwest today. *L. pipixcan* is distinguished from these by a very deep, distinctively outlined brachial depression. Franklin's Gull is a rather scarce migrant in Arizona historically (Monson and Phillips 1981).

Anatidae

Cf. Whistling Swan, *Olor columbianus*. One mandible (SC74) was obtained from a pack rat nest. The species today is a winter resident in Arizona. It will rest on deep water, but feeds where it can reach the bottom with its long neck. We have no comparative osteological material of the Trumpeter Swan, *Olor buccinator*, which is unrecorded for the state in historic times.

Canada Goose, *Branta canadensis*. The four bones from pack rat nests represent apparently three different individuals. The modern status of the various subspecies of Canada Goose wintering in the Southwest is in need of clarification based on specimens (both skins and skeletons). One femur and a radius compare well with the race *B. C. parvipes*, which is known to winter in the state (Phillips et al. 1964), but an ulna (159.5 mm) appears too large for that race.

Snow Goose, *Anser (Chen) caerulescens*. A single bone obtained from a pack rat nest represents this species, which formerly wintered commonly in Arizona.

Mallard, *Anas patyrhynchos*. There are 52 occurrences of this species from the cave. Hargrave considered at least six of these (SC 215, 218, 241, 277, 304, 311) to be in the size range of male *A. p. diazi*, the Mexican Duck. There have undoubtedly been changes in the distribution of this hen-feathered population since Pleistocene times. The Mallard is a common transient and winter resident where there is open water and it breeds locally in suitable habitats in the state (Phillips et al. 1964). This species accounts for 29.5% of the duck bones recovered.

Gadwall, *Anas strepera*. This species is represented by 18 individuals. All appear to be typical of modern Gadwalls except SC 349 which is "conspicuously short and stocky" (Hargrave notes). This duck is a fairly common transient or winter resident in the state, though "in some years this is one of the most abundant ducks wintering along the Colorado River" (Phillips et al. 1964). It represents 10.2% of the ducks recovered in the cave.

Pintail, *Anas acuta*. This species is represented by 18 individuals obtained from pack rat nests. Though in modern times considered "the most abundant Arizona duck" (Phillips et al. 1964) it comprises only 10.2% of the ducks recovered from the cave. Pintails breed in the state at higher elevations, but no immature specimens are represented among the cave material.

Green-winged Teal, *Anas crecca cf. carolinensis*. The species is represented by 11 individuals (6.2% of the recovered ducks). Today this teal is widely distributed and commonly seen in the state as a transient. It "nests rarely on the lakes of Mogollon Plateau" (Phillips et al. 1964) and following an unusually wet winter, nested even in the Lower Sonoran Zone of Maricopa County, Arizona (Bernard Roer, personal communication to Rea).

Blue-winged Teal, *Anas discors*. Three bones were identified as this species, with three others referred (3.4% of the ducks). The Blue-winged is considered the "rarest of the the three teal in Arizona" (Phillips et al. 1964). Hargrave assembled a series of this and the next species, with identifications based on the diagnostic tympanum (auditory bulla).

Cinnamon Teal, *Anas cyanoptera*. Six bones have been identified as this species and one referred (4.0% of the ducks). "The Cinnamon Teal is one of the commonest ducks in Arizona during the spring migration" and is "one of the more numerous ducks breeding in northern Arizona" (Phillips et al. 1964).

Shoveler, *Anas clypeata*. This species is represented by five individuals (2.8% of the ducks). It is a common transient and winter visitant, even on small ponds, though it is seldom seen within the Canyon (Bryan Brown, personal communication).

American Wigeon, *Anas americana*. Thirteen individuals were identified (7.3% of the ducks). One complete tarsometatarsus (SC 232) is immature. This species "breeds occasionally on lakes of the Mogollon Plateau" (Phillips et al. 1964).

Wood Duck, *Aix sponsa*. One scapula (SC 280) compares well with four Wood Ducks in the Hargrave Collection. This is a very rare duck in modern Arizona and unrecorded in Grand Canyon.

Redhead, *Aythya americana*. Nine individuals (6 identified and 3 referred) account for 5.1% of the cave ducks. In historic times the Redhead is more common in Arizona than the Canvasback.

Canvasback, *Aythya valisineria*. This species is represented by 11 individuals (9 identified and 2 referred) or 6.3% of the ducks in the cave. The humerus of this species may be distinguished from *Aythya americana* by the shape of the proximal end (Hargrave notes) as well as by size in males. The Canvasback in modern times is a "rather uncommon migrant and winters in some numbers locally on open waters" (Phillips et al. 1964).

Greater Scaup, *Aythya marila*. One humerus (0.6% of the ducks) was obtained from a pack rat nest. Though the species is represented by only four modern Arizona specimens (collected in December-January), it may be overlooked because of the difficulty of distinguishing it from the more common Lesser Scaup (Phillips et al. 1964).

Lesser Scaup, *Aythya affinis*. Only one specimen (a carpometacarpus, SC 286, measuring 44.5 mm) was identified as this species. In modern times this is by far the most abundant species of *Aythya* in Arizona. Perhaps the poor representation of this scaup is due to the difficulty of identifying bones of the smaller species.

Diving Duck, *Aythya* sp.? Three small individuals (1.7% of the ducks) (SC 197, 227, 228, 229, and 310) of this genus could not be identified to species on characters, but on the basis of size represent either *collaris* or *affinis*.

Cf. Oldsquaw, *Clangula hyemalis*. One partial ulna (SC 194) is referred to this species by Rea. In historic times the Oldsquaw is a rare fall and winter visitant with at least three state specimens (Phillips et al. 1964).

Common Goldeneye, *Bucephala clangula*. Seven individuals, representing 4% of the ducks, have been identified as this species. Today the Goldeneye is considered rare to uncommon in Arizona, but is sometimes locally numerous along the Colorado River (Phillips et al. 1964).

Bufflehead, *Bucephala albeola*. This species is represented in the cave by six individuals (3.4%). This small duck winters in the state and is still a frequent transient (Phillips et al. 1964). All of the bones were recovered from pack rat nests.

Ruddy Duck, *Oxyura jamaicensis*. Four bones from pack rat nests represent probably two individuals (1.1% of the ducks). Though this small duck is common today only in the lower Colorado River valley, it is not infrequently encountered elsewhere in small numbers. "Breeds on Mogollon Plateau, locally along Colorado River, rarely elsewhere" (Phillips et al. 1964).

Common Merganser, *Mergus merganser*. At least four individuals represented by eight bones account for 2.3% of the ducks. This species today is the most common of the three species of merganser in Arizona, nesting here on cliffs. However, all the bones are clearly adult.

Hooded Merganser, *Mergus (= Lophodytes) cucullatus*. One nearly complete ulna (SC 288) was obtained from a pack rat nest. Historically this merganser is an uncommon winter resident, considered by Phillips et al. (1964) to be the rarest of the three species in the state.

Columbidae

Mourning Dove, *Zenaida macroura*. A right humerus from a pack rat nest and a complete femur and proximal end of a tibiotarsus from the 15-20 cm level represents this common species.

Strigidae

Screech Owl, *Otus asio*. A premaxilla and partial mandible were obtained from the 15-20 cm level. They fit into the size range of three modern *O. a. "mycophilus"* (= *aikeni)* skeletons taken from the Grand Canyon which indicates that no size reduction has occurred locally during the Holocene. Modern specimens from the lower Sonoran Desert *(O. a yumanensis, O. a. gilmani)* are smaller.

Great Horned Owl, *Bubo virginianus*. The species is represented by five individuals, one a nestling. This owl probably nested in the cave and could have been responsible for introducing some of the small vertebrate remains via pellets (see Olsen and Olsen, this volume). At least one individual (SC 116), recovered from the trench (25-50 cm), is definitely of Pleistocene age.

Tyrannidae

Black Phoebe, *Sayornis nigricans*. A completely preserved tarsometatarsus (SC 371) was recovered with eleven other bones from the 10-15 cm level. These may have been from an owl pellet. The specimen, 17.05 mm long, is in the size range of a female. It was compared to *Sayornis saya*, which is larger, and *Contopus* and *Empidonax* spp., which are smaller. *"Nuttallornis" (= Contopus) borealis* is shorter and broader.

Say's Phoebe, *Sayornis saya*. A complete humerus was obtained from fill. Both the black phoebe, *Sayornis nigricans*, and the Say's Phoebe are nesting species of the Canyon. The heavier, longer humerus (20.45 mm) indicates it is *S. saya*.

Empidonax Flycatcher, *Empidonax* sp.? A left humerus, with a damaged head, was excavated from the 10-15 cm level (from an owl pellet?). Too few properly prepared reference specimens of the various species were available to make accurate species determination. The characters of the sub-fossil appear nearer *E. hammondii* than *E. difficilis*, but it is closer to the latter in size. It appears not be either *E. wrightii* or *E. traillii*, which both have larger, heavier humeri.

Western Wood Pewee, *Contopus sordidulus*. A complete humerus, excellently preserved, was recovered from 10-15 cm level. It was compared with other small flycatchers such as *Empidonax* and *Sayornis*. Assignment to the species *C. sordidulus* is made on the basis of geographic probability; the closely related (sibling) species, *C. virens*, is a bird of the eastern North American woodlands and is probably osteologically indistinguishable from its western counterpart.

Alaudidae

Horned Lark, *Eremophila alpestris*. A complete humerus (23.4 mm in length) was recovered from the 23-29 cm level. The species has been reported from a number of other Pleistocene deposits (Brodkorb 1978). The size compares well with the races breeding today in Arizona.

Hirundinidae

Swallow sp.? *Hirundo* sp.? Immature bones (partial

sternum, partial pelvis, coracoid, scapula, and miscellaneous vertebrae; surface deposit) appear to be from a non-volant bird, which suggests their nesting in the cave. Although it is referable to *Hirundo*, the Barn Swallow, *H. rustica*, and the Cliff Swallow, *H. lunifrons (pyrrhonota auctorum)*, cannot be distinguished at this early age stage. The closely related Cave Swallow, *H. fulva*, (ranging from southeastern New Mexico to southern Yucatán) cannot be eliminated, though it has never been recorded in Arizona.

Corvidae

Black-billed Magpie, *Pica pica hudsonica*. Four elements (SC 92, 112, 115, 130) from pack rat middens. The Holarctic magpie, represented in North America by the subspecies *hudsonica*, and the Yellow-billed Magpie, *Pica nuttalli* (restricted to California west of the Sierran divide), are undoubtedly cospecific. We have examined skeletons of both forms and have discovered no qualitative differences between them. The Yellow-billed form averages smaller than the Black-billed, though there is a good deal of overlap. The species is represented in Stanton's Cave by a humerus, nearly complete tarsometatarsus, a premaxilla, and a mandible (lacking articulations). Three bones are from the same pack rat nest. The distinctly dark brownish pigment of the rhamphothaca, perhaps foxed after thousands of years, proves conclusively that the Black-billed taxon was involved.

Today the species reaches south only to the Four Corners regions in extreme northeastern Arizona, scarcely enough to qualify as part of the state's avifauna. Its occurrence in the cave, almost two hundred miles west of its modern range, is suggestive of a much better developed riparian condition which this species requires for protection and nesting.

Common Raven, *Corvus corax*. The distal end of the a right tibiotarsus (SC 84) was recovered from the surface of the tunnel to Antechamber One. The small size of the bone (greatest width across condyles, anterior view, 10.7 mm) indicates a very small female of the race *C. c. sinuatus*, currently breeding in the Southwest (the most likely probability) or a male of the Southern California and Baja California race *C. c. clarionensis* (see map, page 200, in Rea 1983). The bone is nevertheless much larger than the White-necked Raven, *C. cryptoleucus*, breeding in the desert grasslands of the Southwest.

The distal end of a humerus (SC 83) with an attached partial radius is the size of a female *C. c. sinuatus*, but larger than a male *C. c. clarionensis*. The greatest width of the cave humerus is 18.7 mm. All three bones were obtained from the same area and may represent a single individual.

Cf. Scrub Jay, *Aphelocoma coerulescens*. A femur, damaged humerus, and partial pelvis (SC 372-374) were obtained from surface deposit. Assuming that these three elements are from the same individual, the proportions indicate *Aphelocoma coerulescens* rather than *Cyanocitta stelleri*, the Steller's Jay, also present in Grand Canyon. Characters of the femur distinguish it from the closely related Piñon Jay, *Gymnorhinus cyanocephalus*.

Laniidae

Northern Shrike, *Lanius excubitor*. A complete humerus (SC 368) was obtained on the talus slope inside the cave at the entrance. The relatively good preservation indicates the bone was of quite recent deposition (surely not more than several years). Although *L. excubitor* and *L. ludovidianus*, Loggerhead Shrike, are rather similar in size and very close in color and pattern, the stoutness of the shaft, width of the distal end, and width of the bicipital crest clearly rule out *L. ludovidianus*, the species common in the Southwest. There are several winter records for the northern shrike, though it probably occurs more commonly in the northern counties of the state (Phillips et al. 1964). This appears to be the first specimen record for the Grand Canyon (Bryan Brown, personal communication).

The family Laniidae is placed here, following the Corvidae which it most closely resembles anatomically, rather than farther along in the passerine sequence, where it is usually found in current checklists.

Cinclidae

American Dipper, *Cinclus mexicanus*. A right humerus (SC 375) (proximal end chipped) was obtained from the 5-30 cm horizon of the main room grid. The species is resident in Grand Canyon.

Troglodytidae

Canyon Wren, *Catherpes mexicanus*. One rectrix was obtained from 5-10 cm level. This tail feather appears to be very recent and may have been dropped by a bird foraging within the cave during the course of excavation. Canyon Wrens are partial to rugged country, where they forage deep beneath rocks and in crevices.

Rock Wren, *Salpinctes obsoletus*. A complete pelvis with caudal vertebrae was recovered from a pack rat nest. This species and the Canyon Wren are both common in the Canyon bottom today.

Turdidae

Northern (Common) Robin, *Turdus migratorius*. One complete femur (pack rat midden near FS #3) is definitely this species, not *T. grayi*.

Clay-colored Robin, *Turdus grayi*. In a pack rat midden adjacent to FS #13 was obtained a virtually complete skull with attached mandible, quadrates, os opticus (one only), basal plate (cricoid) of laryngeal structure, and medial portions of hyoid apparatus (Figure 7-1 and 7-2). The skull has suffered some damage (rodent gnawing?) to the basitemporal plate and auditory bullae, and part of the cranium is missing; it is an adult based on its pneumaticity. Some strands of muscle are still attached to the left orbital process of the quadrate and some feather fragments (not *in situ)* adhere to the cranium.

Turdid genera such as *Zoothera* (= *Hesperocichla), Hylocichla, Catharus,* and even South America *Platatycichla* were readily eliminated, leaving the skull unquestionably a *Turdus* of some species. The only United States species is the abundant and widespread Northern (Common) Robin, *T. migratorius.* However, comparison with approximately 75 skulls of this species (AMR, ARIZ, MVZ, SD) failed to reveal one with similar characters. Since the correct determination of this fossil is so critical, we include the following notes, leading to our conclusion. (Comparisons were also made with other New World species which proved to be less closely related.)

The fossil differs from *T. migratorius* in: 1) ventral profile of ectethmoid plate entirely different in size and shape, being narrow and extending well beyond jugal bars laterally; 2) palatines expanding more posteriorly; 3) breadth of quadrate greater; 4) mandibular foramen considerably smaller. The only other *Turdus* species besides *T. migratorius* occurring today in the Southwest (merely as a vagrant) is *T. rufopalliatus*, the Rufous-backed Robin of western México. The fossil is distinguishable from this species by: 1) palatines posteriorly *ca.* two times broader and differently shaped; 2) vomer apparently stouter; 3) internal process of mandible with a different configuration. *Turdus assimilis*, White-throated Robin, another widespread Mexican species that ranges as far north as southeastern Sonora, is similar to the fossil in many ways, but has more lacrimal area and base to the ectethmoid plate. However, in all these characters *Turdus grayi* fits the fossil exactly.

The Clay-colored Robin occurs on the gulf slope of México from Nuevo León and Tamaulipas south and east to the Yucatán Peninsula and Chiapas (Figure 7-3). It has been seen (as an accidental) in extreme southern Texas (Brownsville and Mission, Peterson 1961). It is not so unlikely that this species ranged naturally to Arizona sometime in that last 40 to 80 thousand years when more warm and mesic conditions prevailed in the Southwest.

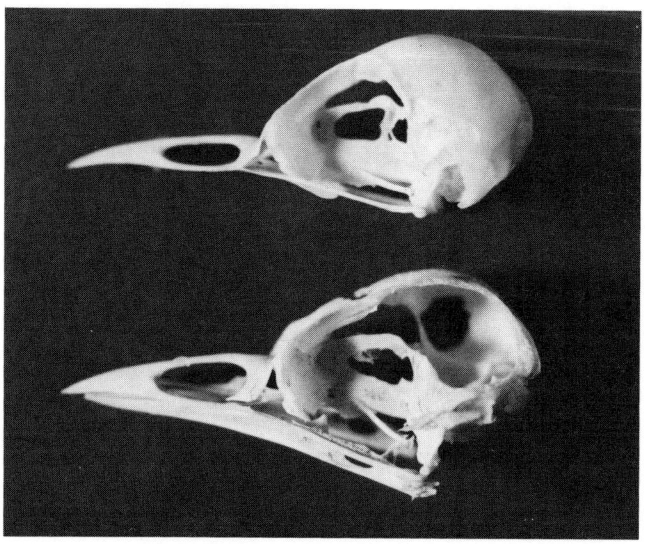

Figure 7-1. Lateral view of robin crania. (above) *Turdus migratorius* (without mandible); (below) Stanton's Cave skull referred to *T. grayi* (mandible intact).

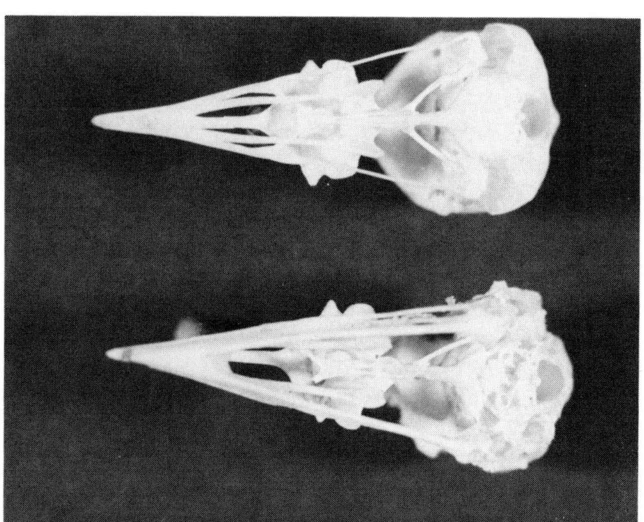

Figure 7-2. Palatal view of robin skulls. (above) *Turdus migratorius* (without mandible); (below) Stanton's Cave *T. grayi*.

Figure 7-3. Turkey and robin distribution. The extinct turkey, *Meleagris crassipes*, described from Nuevo Leon, is now known from ten fossil sites (triangles), including Stanton's Cave (barred line; after Rea 1980). The Clay-colored Robin, *Turdus grayi* (crosshatching) ranges north today to Nuevo Leon and Tamaulipas.

Mountain Bluebird, *Sailia currucoides.* A complete ulna, chipped carpometacarpus, and referred coracoid were obtained from the 15-20 cm level of Grid FF. The male of the Western Bluebird, also occurring today in the Canyon, is considerably smaller. This appears to be the first record of this species for the early Holocene.

Townsend's Solitaire, *Myadestes townsendi.* The species is represented by seven attached wing bones distal to the humerus, including a complete carpometacarpus, recovered from the 10-15 cm level. Although the family Turdidae is of worldwide distribution, the solitaires are predominantly a montane group from the southern part of North America, northern South America, and the West Indies. *M. townsendi* is the only species to penetrate into the north — as far as Alaska and the southern Yukon. There appears to be no other fossil record for this species. An extrapolated date for the 10-15 cm level of the cave is approximately 5+ thousand years B.P. The species is common today in juniper woodlands of northern Arizona.

Parulidae

Yellow-rumped Warbler, *Dendroica coronata.* A complete carpometacarpus and tibiotarsus were obtained from the 15-20 cm level of Grid EE. The species breeds commonly in the Transition and Canadian life zones on the Canyon rims and is abundant migrant throughout the state. The specimens were compared with other small passerines that might be expected in the Southwest such as Troglodytidae, Sylviidae, Aegithalidae, Remizidae, Paridae, as well as other species and genera of Parulidae. The characters, absolute size, and relative proportions fit only this species. Whether these bones belong to the Myrtle subspecies group or the Audubon's subspecies group of the Yellow-rumped complex cannot be determined osteologically.

Icteridae

Northern Oriole, *Icterus galbula.* A left ramus of a mandible was recovered from the 15-20 cm level (early post-glacial) of Grid FF. The bone was compared also with *I. parisorum* and *I. cucullatus* (more slender-billed species breeding in the state today), and with the Mexican species *I. pustulatus, I. gularis,* and *I. pectoralis,* which are heavier billed. The fossil differs from all of these in characters as well as size. The two western subspecies of *I. galbula* were formerly known as Bullock's Oriole.

Emberizidae

A number of well-preserved bones representing the finch-sparrow assemblage remain to be identified. These bones from Stanton's Cave will no doubt add greatly to the meager knowledge of emberizids in the Pleistocene.

Bunting, *Passerina* sp. A complete left humerus (SC 370) and apparently matching head of a right humerus were recovered from the 10-15 cm level in a cluster of 12 bones apparently representing a raptor pellet. The excellently preserved humerus appears indistinguishable from the Varied Bunting, *P. versicolor* (two males compared), but it may be a male Indigo Bunting, *P. c. cyanea*, or a female Lazuli Bunting, *P. c. amoena*, the most probable on geographic grounds. It is smaller than male Lazuli and smaller and less bulbous proximally than the Painted Bunting, *P. ciris.*

Dark-eyed Junco, *Junco hyemalis.* This species is represented from two difference proveniences: a femur, ulna, and two partial scapulae from Grid FF, 20-25 cm horizon, and an ulna from Grid DD, 15-40 cm horizon. These bones are all larger and heavier than such northern and western junco races as *J. h. mearnsi, J. h. simillimus, J. h. thurberi,* and *J. h. shufeldti* which winter abundantly in the Southwest. They agree in size with the Gray-headed Junco, *J. h. caniceps,* suggesting that there has been no change in junco size since the late Pleistocene or early Holocene. No skeletons were available of the local breeding race, *J. h. dorsalis,* the other member of the Gray-headed group.

White-crowned Sparrow, *Zonotrichia* cf. *leucophrys.* A humerus, associated head of a coracoid, and distal end of an ulna were recovered from the 10-15 cm level. The humerus is similar to, but distinguishable from, humeri of *Junco* and *"Melospiza"* (= *Passerella) melodia.* It is apparently indistinguishable from humeri of *Z. albicollis,* but this species is unlikely on geographic grounds. In winter *Z. leucophrys* is one of the most abundant and widely distributed emberizids in the Southwest.

Fox Sparrow, *Passerella iliaca.* A mandible (SC 366) missing most of the right ramus, was recovered from the 0.1 cm level. This mandible is only moderately swollen, hence about the size of one of the northern or interior races of Fox Sparrow which winter in Arizona, rather than one of the southern Pacific races with more massive bills, which seldom reach the Southwest in winter.

Discussion

Bird bones no doubt accumulated in Stanton's Cave in a number of ways. Some species may have entered the cave voluntarily for nesting or shelter. These would include scavengers, certain raptors (hawks and owls), and such passerines as swallows and wrens that use natural cavities. We identified young of *Cathartes aura, Aquila chrysaetos, Falco mexicanus, Falco sparverius, Bubo virginianus,* and some passerine species.

Mammalian predators may have been responsible for introducing some of the birds as prey items. This may explain the large number of ducks, which would not have entered the cave of their own accord. Olsen and Olsen (this volume) report the following predatory mammals from Stanton's Cave: coyote, *Canis latrans;* gray fox, *Urocyon cinereoargentus;* ringtail; *Bassariscus astutus;* raccoon, *Procyon lotor* (one individual). Otters *(Lutia canadensis)* may have brought much of the fish and duck materials into the cave. We noted mammal tooth marks in addition to the characteristic rodent gnawing on a number of duck wings. Water dependent species accounted for 58.8% of the birds reported here, with 52.9% of the macro-avifauna being ducks. It should be noted, however, that high proportions of aquatic birds occur in virtually all other Southwest and Great Basin cave deposits of Pleistocene age.

Hawks and owls no doubt brought other birds into the cave to feed their young and raptors regurgitated prey items as pellets while roosting in the cave. Even scavenging birds (Vulturidae [= Cathartidae]) could have introduced vertebrate bones in this manner (Rea 1973). Young of ledge-nesting birds may have fallen to their death near the cave mouth. Pack rats undoubtedly gathered bird bones along with plant materials about the entrance to deposit in their middens. There is no evidence of humans bringing food items into the cave.

Compared to southern New Mexico cave avifaunas, Stanton's Cave contains surprisingly few paleospecies. These include a teratorn, an enormous bird related to New World vultures and storks, now believed to be only secondarily a scavenger (Campbell and Tonni 1981), the Ancestral Condor, believed by us to be only a temporal species of the California Condor, and the extinct turkey. Missing from the fauna are the various extinct Pleistocene raptors, falconid and accipitrid scavengers, and the vulture *Coragyps*, all well represented in California asphalt deposits and New Mexico dry caves.

Two species, *Centrocercus urophasianus* and *Pica pica*, have been locally extirpated. The demise of the Sage Grouse is correlated with a decline in *Atriplex* as a result of the dryer, warmer local conditions in the postglacial and overgrazing historically. The Mt. Trumbull region sight record might legitimately be questioned, but if the species persisted there, it was only as a relictual population. The historic restriction of the Black-billed Magpie to the Four Corners region is related to the decline of developed riparian woodlands with cottonwoods. There are some recent sight records from throughout the Grand Canyon area west to Lake Mead (Monson and Phillips 1981).

Eight species were identifiable to subspecies, or at least subspecies group, on the basis of size or, in one specimen, bill color: *Cathartes aura, Falco sparverius, Otus asio, Eremophila alpestris, Pica pica, Corvus cora, Junco hyemalis, Passerella iliaca*. In each case the infraspecific form recovered is the expected one for the area today, suggesting size stability in the Holocene.

Condor bones from the surface of this and other caves in the Grand Canyon are particularly interesting, since early field ornithologists reported the species from western and northwestern Arizona (summarized in Rea 1981). It is possible that the species survived there in diminished numbers into historic times. P.S. Martin (personal communication to Rea) believes this population did not survive the local extinction of the megafauna about 11 thousand years B.P. Surface bones from various caves in the Canyon should be directly radiocarbon dated in an effort to resolve this problem. Rea (1981) has proposed the Grand Canyon as a safe and suitable area for the reintroduction of captive bred condors.

Seven neospecies and the genus *Passerina* are reported here apparently for the first time from Pleistocene/Holocene contexts (based on Brodkorb's fossil catalogs): *Buteogallus anthracinus, Larus pipixcan, Sayornis nigricans, Contopus sordidulus, Turdus grayi, Sialia currucoides, Myadestes townsendi*. The Black Hawk, Franklin's Gull, and Clay-colored Robin are from undated contexts. The two flycatchers, the solitaire, and the bunting are from the 10-15 cm level which is extrapolated to an age of 5+ thousand years B.P. The bluebird is from the 15-20 cm postglacial stratum.

The only unexpected recoveries were the extinct turkey, *Meleagris crassipes*, and the Clay-colored Robin, *Turdus grayi*, both species with Caribbean Mexican affinities (Figure 7-3). The turkey, originally described from San Josecito Cave in Nuevo León (Miller 1943) and long known only from the local fauna, has recently been identified from nine additional sites, all in the Southwest (Rea 1980). Steadman (1980) has traced the entire paleontological history of the Meleagridinae. There is no good evidence for sympatry of this distinctive form with the surviving *M. gallopavo*, though the two species were contemporaneous in the late Pleistocene. *Teratornis merriami* was recovered likewise from both Stanton's Cave and San Josecito Cave, as well as from the California asphalt deposits, but these enormous birds must have ranged broadly across the continent. Today the Clay-colored Robin breeds northward on the Caribbean slope to central Tamaulipas and southern central Nuevo León. These associations cannot be explained on the basis of the usual model of a postglacial northward or vertical displacement of montane-woodland species (see discussion in Mead 1981). Similarly, the gray-breasted crake, *Laterallus exilis*, ranging today in northeastern South America with stragglers north in Central America to Belize (Russell 1966; see map in Olson 1974:171), was tentatively identified by Storrs L. Olson from a Paleo-Indian site in western Texas (Lubbock Lake; Rea, unpublished MS). Since at least two avian species with subtropical Caribbean affinities were recovered in the Grand Canyon local fauna, this biogeographic association should be considered for other groups of organisms as well.

Acknowledgments

Dr. R. Roy Johnson commented on an earlier draft of the manuscript. Drs. David W. Steadman, Jim I. Mead, and Thomas R. Van Devender reviewed a later version. Their helpful comments are appreciated. We thank Mr. Robert M. Chandler for the fossil photographs. We are especially appreciative of Dr. Hildegarde Howard for her determinations of the condors and teratorns. We thank Dr. Paul W. Parmalee for assistance with the grouse identifications and for loaning critical comparative materials. Before his death in 1978, Dr. Lyndon L. Hargrave read and approved his contributions on the other large bird bones from the cave.

Table 7-1. Occurrences of bird bones in Stanton's Cave.

Species[1]	Post-Pleistocene (0-20 cm)[2]	Pleistocene (20+ cm)	Midden (undated)	Modern Occurrence[4]
Podicipeps nigricollis	1[3]		1	+[5]
Aechmorphorus occidentalis			2	(+)
Teratornis merriami		1	1	0
Cathartes aura	1		2	+
Gymnogyps (amplus-sized)	2 (+2?)	5	4	0
Gymnogyps (amplus or californianus)	4	2	4	0
Gymnogyps (unassignable)	5	4	3	0
Ardea herodia			3	+
Accipiter striatus	1			+
Buteo jamaicensis			2	+
Buteogallus anthracinus			2	0
Aquila chrysaetos	3		1	+
Circus cyaneus			1	(+)
Falco mexicanus	2		1	+
Falco sparverius	2	1		+
Centrocercus urophasianus	1 (+?)			
Gallinula chloropus			1	0
Fulica americana	2		5	+
Capella gallinago		1		+
Numenius americanus			1	(+)
Actitis macularia		1		+
Erolia melanotos		1		0
Phalaropus fulicarius		1		0
Lobipes lobatus		2		(+)
Recurvirostra americana			1	+
Larus pipixcan	1		1	0
Olor cf. columbianus			1	(+)
Branta canadensis			4	(+)
Anser caerulescens			1	(+)
Anas platyrhynchos	12	1	39	+
Anas strepera	1		17	(+)
Anas acuta			18	+
Anas crecca	4		7	+
Anas discors	2	1	3	+
Anas cyanoptera	2	0	5	+
Anas clypeata	1		4	(+)
Anas americana	3		10	+
Aix sponsa			1	0
Aythya americana	3		6	(+)
Aythya valisineria	1		10	(+)
Aythya marila			1	0
Aythya affinis			1	(+)
Aythya (cf. affinis or collaris)	1		2	+
Cf. Clangula hyemalis			1	0
Bucephala clangula	1		6	+
Bucephala albeola			6	+
Oxyura jamaicensis			2	+
Mergus merganser			4	+
Mergus cucullatus			1	(+)
Zenaida macroura	1		1	+
Otus asio	1			+
Bubo virginianus	2	1	3	+
Sayornis nigricans	1			+
Sayornis saya	1			+

Species[1]	Post-Pleistocene (0-20 cm)[2]	Pleistocene (20+ cm)	Midden (undated)	Modern Occurrence[4]
Empidonax sp.	1			+
Contopus sordidulus	1			+
Eremophila alpestris	1			+
Hirundo (rustica or ?)	1			(+)
Pica pica hudsonica			4	0
Corvus corax	1 (2?)			+
Cf. Aphelocoma coerulescens	1			+
Lanius excubitor	1			(+)
Cinclus mexicanus		1		+
Catherpes mexicanus	1			+
Salpinctes obsoletus	1			+
Turdus migratorius			1	+
Turdus grayi			1	0
Sialia currucoides	1			+
Myadestes townsendi	1			+
Dendroica coronata	1			+
Icterus galbula	1			(+)
Passerina sp.?	1			+
Junco hyemalis		2		+
Zonotrichia cf. leucophrys	1			+
Passerella iliaca	1			(+)

[1]Nomenclature follows the American Ornithologists' Union Check list, 1957, its 1973 supplement, Phillips et al. 1964, and Rea 1983. See text for English names.
[2]Including surface bones.
[3]Numbers indicate presumed MNI (Minimum Number of Individuals) rather than total number of bones recovered.
[4]Based on Grater (1937), Brown et al. (1978), and Bryan Brown (personal communication).
[5]0 = absent from region, + = present and expected, (+) = occasional or very rare.

Footnote

1. Where no SC# is given, for some reason the late Dr. Hargrave did not assign a number.

References

American Ornithologists' Union
1957 *Check list of North American birds.* 5th ed. Baltimore: City Port Press, Inc.

American Ornithologists' Union.
1973 Thirty-second supplement to the American Ornitholgists' Union checklist of North American birds. *Auk* 90:411-419.

Behle, W.H., and M.L. Perry
1975 *Utah birds: check list, seasonal and ecological occurrence charts and guide to bird finding.* Salt Lake City: Utah Museum of Natural History.

Brodkorb, P.
1964 *Catalogue of fossil birds,* part 2 (Anseriformes through Galliformes). Bull. Florida State Museum, Biol. Sci. 8:195-335.
1978 *Catalogue of fossil birds,* part 5 (Passeriformes). Bull. Florida State Museum, Biol. Sci. 23:139-228.

Brown, B. T., and others
1978 *Birds of the Grand Canyon, an annotated checklist.* Grand Canyon Natural History Association, Monograph 1.

Campbell, K.E.
1979 *The non-passerine Pleistocene avifauna of the Talara Tar Seeps, northwestern Peru.* Life Sci. Contrib. Royal Ontario Museum No. 118.

Campbell, K.E., and E.P. Tonni
1981 Preliminary observations on the paleobiology and evolution of teratorns (Aves: Teratornithidae). *J. Vert. Paleontology* 1:265-272.

Carothers, S.W., and R.R. Johnson
1975 Recent observations on the status and distribution of some birds of the Grand Canyon region. *Plateau* 47:140-153.

deSaussure, R.
1956 Remains of the California Condor in Arizona caves. *Plateau* 29:44-45.

Euler, R.C.
1978 Archeological and Paleobiological Studies at Stanton's Cave, Grand Canyon National Park, Arizona — A Report of Progress. *National Geographic Society Research Reports, 1969 Projects* 141-162.

Fisher, H.I.
1944 The skulls of cathartid vultures. *Condor* 46: 272-296.
1947 The skeletons of recent and fossil Gymnogyps. *Pacific Sci.* 1:227-236.

Grater, R.K.
1937 *Check List of birds of Grand Canyon National Park.* Natural History Bull. No. 8. Grand Canyon Natural History Association.

Hargrave, L.L.
1970 *Feathers from Sand Dune Cave: a Basketmaker cave near Navajo Mountain, Utah.* Technical Series No. 9, Museum of Northern Arizona.

Howard, H.
1952 The prehistoric avifauna of Smith Creek Cave, Nevada, with a description of a new gigantic raptor. *Bull. Southern Calif. Aca. Sci.* 51:50-54.
1962 A comparison of avian assemblages from individual pits at Rancho La Brea, California. *Los Angeles Co. Museum Contr. Sci.* 58:1-24.
1963 Fossil birds from the Anza-Borrego Desert. *Los Angeles Co. Museum Contr. Sci.* 73:1-33.
1972 The incredible teratorn again. *Condor* 74:341-344.

Huey, L.M.
1939 Birds of the Mount Trumbull region, Arizona. *Auk* 56:320-325.

Johnsgard, P.A.
1973 *Grouse and quails of North America.* Lincoln: Univ. Nebraska Press, .

King, J.E.
1973 Modern pollen in the Grand Canyon, Arizona. *Geoscience and Man.* 7:73-81.

Koford, C.B.
1966 *The California Condor.* Dover, N.Y.

Martin, P.S., B.E. Sables, and R. Shutler, Jr.
1961 Rampart Cave coprolite and ecology of the Shasta ground sloth. *Amer. Jour. Sci.* 259:102-127.

Mead, J.I., and A.M. Phillips, III
1981 The late Pleistocene and Holocene fauna and flora of Vulture Cave, Grand Canyon, Arizona. *Southwestern Naturalist* 26:257-288.

Miller, L.H.
1943 The Pleistocene birds of San Josecito Cavern, Mexico. *Univ. Calif. Publ. Zool.* 47:143-168.
1960a Condor remains from Rampart Cave, Arizona. *Condor* 62:70.
1960b On the history of the Cathartidae in North America. *Novadades Colombianas* 1:232-235.

Monson, G., and A.R. Phillips
1981 *Annotated checklist of the birds of Arizona,* second ed. Tucson: Univ. Arizona Press.

Olsen, J.W. and S.J. Olsen
1984 "Zooarchaeological analysis of small vertebrates from Stanton's Cave, Arizona." This volume.

Olson, S.L.
1974 The Pleistocene rails of North America. *Condor* 76:169-175.

Parmalee, P.W.
1969 California Condor and other birds from Stanton Cave, Arizona. *J. Arizona Acad. Sci.* 5:204-206.

Peterson, R.T.
1961 *A field guide to western birds.* Cambridge: Riverside Press.

Phillips, A., J. Marshall, and G. Monson
1964 *The birds of Arizona. Tucson: Univ. Arizona Press.*

Phillips, A.R.
1946 *The birds of Arizona.* Doctoral dissertation submitted to Cornell University, N.Y.
1968 "The instability of the distribution of land birds in the Southwest." *In* A.H. Schroeder (ed.), *Collected papers in honor of Lyndon Lane Hargrave.* Papers Archaeological Soc. New Mexico 1:129-162.

Rea, A.M.
1973 Turkey Vultures casting pellets. *Auk* 90:209-210.
1980 "Late Pleistocene and Holocene turkeys in the Southwest." *In* K.E. Campbell, Jr. (ed.), *Papers in avian paleontology honoring Hildegard Howard.* Los Angeles Co. Mus. Contrib. in Sci. No. 330:209-224.
1981 California Condor captive breeding: a recovery proposal. *Environment Southwest* No. 492:8-12.
1983 *Once a river: bird life and habitat changes on the middle Gila.* Tucson: Univ. of Arizona Press.

Russell, S.M.
1966 Status of the Black Rail and the Gray-breasted Crake in British Honduras. *Condor* 68:105-107.

Steadman, D.W.
1980 "A review of the osteology and paleontology of turkeys (Aves: Meleagridinae)." *In* K.E. Campbell, Jr. (ed.), *Papers in avian paleontology honoring Hildegard Howard.* Los Angeles Co. Mus. Contrib. in Sci. No. 330:131-207.

Wetmore, A.
1935 The Mexican Turkey Vulture in the United States. *Condor* 37:176.

Wetmore, A., and H. Friedmann
1933 The California Condor in Texas. *Condor* 35: 37-38.

Chapter 8

Dendrochronology of Driftwood from Stanton's Cave

by

C. W. Ferguson
Laboratory of Tree-Ring Research
The University of Arizona, Tucson

In my search for old wood for use in the calibration of the radiocarbon time scale beyond the 7104-year tree-ring chronology then available for bristlecone pine *(Pinus longaeva)* (Ferguson 1969), my attention was drawn to an extensive deposit of driftwood in Stanton's Cave, at Mile 31.7 (51 river kilometers below Lees Ferry, Arizona) in the Grand Canyon.

The initial discovery of driftwood in the two trenches excavated in the cave (Euler 1978) disclosed an unexpected amount of coniferous wood as contrasted to more obvious large-sized logs of cottonwood *(Populus* sp.). Preparation of a selected few logs of Douglas-fir *(Pseudotsuga menziesii)* disclosed two specimens with ring sequences of greater than average length. Of these, one (STC-1) of the two best (with STC-2; see Figures 8-1 and 8-2) was selected for radiocarbon analysis.

Based upon the 4095-year radiocarbon age of a split-twig figurine found on the surface of the cave floor (Euler and Olson 1965; Euler 1966), and the depth and character of the deposits, it was felt that the underlying wood was deposited on what then was the cave floor in the range of 12,000 years ago.

However, the initial driftwood specimen, collected in the 1969 excavation, gave the rather surprising ^{14}C range of greater than 35,000 years (University of Arizona A-1056; t[½] = 5568).[1] This date, much too early to be of value in the ^{14}C calibration studies, resulted in a change in emphasis. The major objective in the dendrochronological study of wood from Stanton's Cave itself became to prove or disprove the contemporaneity of the deposit. Some crossdating was found in the tree-ring chronologies of separate specimens, but units of two or more crossdated specimens could not be matched with each other, indicating a possible spread in time for deposition in the cave.

The mouth of Stanton's Cave is 144 feet (44 meters) above the present level of the Colorado River. How this cave became filled with driftwood is the subject of another paper in this volume (Hereford). This led to the idea that jams of driftwood elsewhere along the river might contain deposits of more recent age, and that a collection of available contemporary driftwood would permit us to learn something of species, site relationships, and sources of origin for the period predating the construction of Glen Canyon Dam.

A group of specimens from the cave, selected for their species (Douglas-fir), length of series, and quality of record was more intensely studied. Ring sequences

Figure 8-1. Cross section of Douglas-fir log (STC-1) selected for radiocarbon analysis. The ten-year interval chosen is indicated by "C-14" mark near the eroded center; a five-centimeter rule provides a relative scale.

along one or more radii were measured and plotted. The plots and ultimately the wood were checked for possible crossdating. Numbers of specimens and years of growth of each are presented as a histogram (Figure 8-3) and show an unexpected curve with a greater proportion of shorter series.

The numbers of rings per specimen, which would be adequate for undisputed crossdating, is unfortunately counterbalanced by the poor quality of the ring sequences. The relatively poor sensitivity evidently reflects the riparian site conditions under which the trees grew. Despite these adverse qualities, a few of the specimens did contain ring series which apparently could be crossdated.

In addition to cottonwood and Douglas-fir, a few other species were represented in the Stanton's Cave driftwood. These included *Juniperus* sp. and what probably is a legume.

The provenience of 40 Douglas-fir specimens, tabulated in Table 8-1, shows a concentration in units O and P of the North-South Trench. These probably represent a collection of smaller pieces trapped in back of the large cottonwood logs shown in the trench profile.

Considering that some of the hundred-plus pieces collected were exploratory as to species, the percentage of dendrochronologically dated specimens indicate that the approach — using driftwood to date or interpret events in the Grand Canyon — is feasible. The 957-year period spanned by the tree-ring series in 21 pieces of contemporary driftwood (Ferguson 1971) pro-

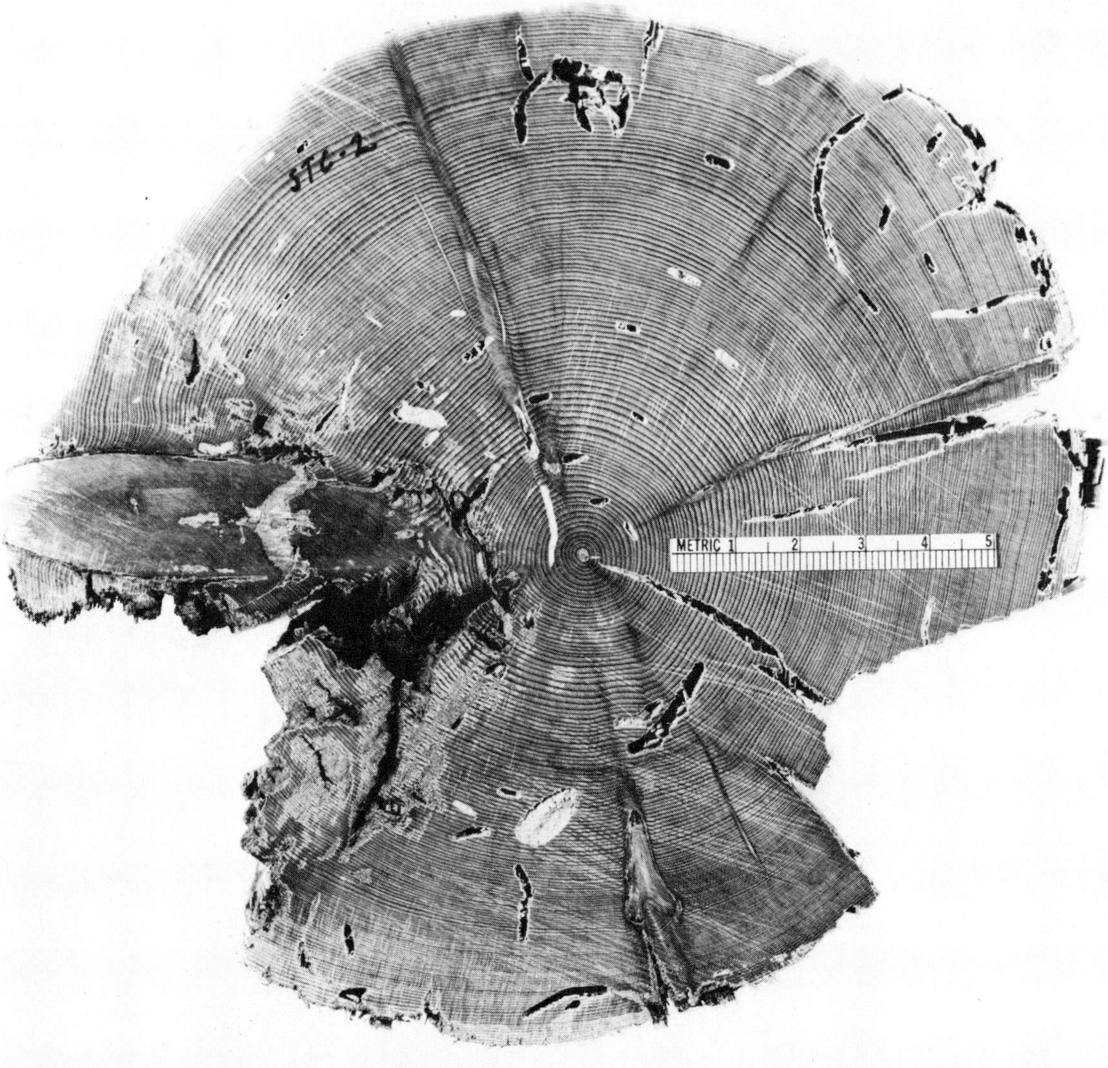

Figure 8-2. Cross section of Douglas-fir log (STC-2).

vides a basis for interpreting the scattering of tree-ring sequences through time, with a general grouping of specimens in the three intervals A.D. 1300-1600, 1600-1800, and 1830-1940, indicated that two or more specimens that crossdate with each other may not crossdate with other such units.

A piece of piñon wood, about 30 cm long and 6 cm in diameter, was found at the upper end of the mapped area. A median section was removed from the remnant, designated STC-3. The measured and plotted radius contained 136 rings. Because the specimen and its provenience represented certain anomalous features and because its ring sequence was potentially datable, it was selected for radiocarbon analysis. A gross unit, without reference to specific rings, was prepared for radiocarbon analysis. The specimen was selected for ^{14}C analysis because of the species; *Pinus edulis* was barely represented in the driftwood deposit in the cave, although it is relatively common along the river today. The original bulk specimen had a coating of cave dust, but contained some apparently recent breaks. This may indicate a recent redeposition, as by Indians or early river parties. Even though the wood was of a species and size that made it highly combustible, it was not of the form commonly used as cave torches.

The ring sequence is of a quality that would enable the specimen to be dated against the Southwest archaeological chronology. No such effort was made prior to the ^{14}C age determination because of the assumed great age of the specimen. The earliest ring on contemporary driftwood is A.D. 1011. The second oldest is a 630-year sequence dating back to A.D. 1042; its innermost ring is a function of the long life of the tree, not time since death. The radiocarbon date of 1500± B.P. (A1184) places the specimen in the range of A.D. 450, a time period in which the limited control chronologies have not made possible the dendrochronological dating of this specimen.

Based upon my study of the wood from the cave and of present-day driftwood, I feel that the wood in Stanton's Cave represents a single, short-term depositional time period, perhaps even one year; certainly no more than a few.

Acknowledgments

Initial and continuing interest in dating driftwood and my participation in the 1970 cave excavation came from the Stanton's Cave study, an interdisciplinary project conducted by Robert C. Euler, then of Prescott College, under terms of a grant from the National Geographic Society for the "Paleoclimatic History of the Grand Canyon." The exploratory collection of contemporary driftwood was made on a 226-mile river trip, sponsored by the Arizona Academy of Science, 18-26 August 1970. Partial financial support for the river trip was provided by the Laboratory of Tree-Ring Research.

Figure 8-3. A histogram showing the distribution, by ten-year units, of the length of the individual measured tree-ring series.

Table 8-1. Provenience of 40 pieces of Douglas-fir which were measured and plotted.

	Grid	Depth (cm)	Specimen	Number of Specimens
East-West Trench:				
	FF	25-50	132	1
	GG	50-75	166	1
	HH	50-75	154, 161	2
North-South Trench:				
	O-19	40-65	17, 63, 64, 71A, 71B 78, 96, 105, 108	9
		65-90	1, 113, 114, 116, 188, 189	6
		95-105	212	1
	P-19	35-60	AD, EF, BH, CG, 21, 25, 31, 138, 144, 145, 148, 169, 175, 176	14
		60-85	200	1
	Q-19	35-60	53, 136, 221	3
Other:				
SE Quadrant-surface			184	1
Lost Provenience			990	1
			Total	40

Footnote

1. A more recent ^{14}C date, derived since this was written, from a piece of the same log as that dated at the University of Arizona, is 43,700 + 1800 and -1500 years B.P. (U.S.G.S. No. 1774) analyzed at the radiocarbon laboratory of the U.S. Geological Survey at Menlo Park, California, by S.W. Robinson (personal communication, 1984) — Editor.

References

Euler, Robert C.

1966 Willow figurines from Arizona. *Natural History* 75:3:62-67.

1978 Archeological and Paleobiological Studies at Stanton's Cave, Grand Canyon National Park, Arizona — A Report of Progress. *National Geographic Society Research Reports, 1969 Projects* 141-162.

Euler, Robert C., and Alan P. Olson

1965 Split-twig figurines from Northern Arizona: new radiocarbon dates. *Science* 148:3668:368-369.

Ferguson, C.W.

1969 A 7104-year annual tree-ring chronology for bristlecone pine *(Pinus aristata)*, from the White Mountains, California. *Tree-Ring Bulletin* 29:3-4:1-29.

1971 Tree-ring dating of Colorado River driftwood in the Grand Canyon. *Hydrology and Water Resources in Arizona and Water Resources in Arizona and the Southwest,* 1:351-366. Proceedings of the 1971 Meetings of the Arizona Section — American Water Resources Association and the Hydrology Section, Arizona Academy of Science, Tempe. Tucson: University of Arizona Press.

Chapter 9

Driftwood in Stanton's Cave: The Case for Temporary Damming of the Colorado River at Nankoweap Creek in Marble Canyon Grand Canyon National Park , Arizona[1]

by

Richard Hereford
U.S. Geological Survey
Flagstaff, Arizona

Abstract

Driftwood in Stanton's Cave in Grand Canyon National Park, Arizona, has been dated at greater than 35,000 years[2] and may have floated into the cave on a lake that formed behind a dam in the Nankoweap Creek area of Marble Canyon. The dam probably resulted from a rockfall that originated from the east wall of Marble Canyon. Debris from the rockfall is preserved at the mouth of Nankoweap Creek. The reconstructed area of the deposit indicates that it was large enough to block the Colorado River, and a gravel deposit at the top of the debris indicates that the river flowed over the rockfall material. If the rockfall and driftwood are of the same age, a reasonable assumption with the information available, then ponding of the Colorado River by the rockfall was the event that flooded Stanton's Cave and deposited the driftwood.

Introduction

In a progress report of research in Stanton's Cave in Grand Canyon National Park, Arizona, Euler (1978:156) noted that driftwood resting on the bedrock floor of the cave was deposited more than 35,000 years[2] ago; a sample of the wood was radiocarbon dated to that time period (A1056-UA; Euler 1978:159). The wood occurs at the base of the cave deposits, which are 50 to 80 cm in thickness and consist of gray-brown clayey silt overlain by 5 to 15 cm of dark organic material. It is not clear whether the clayey silt was deposited with the driftwood or if it was washed into the cave from the rear. The available evidence indicates that the cave was not repeatedly flooded by the Colorado River.

The wood, at an elevation of 927 meters, which is 44 meters above the river, far above the level of historic floods, drifted into the cave during high water caused by temporary ponding downstream, or by exceptionally high discharge (Euler 1978). This report explores each possibility and suggests that a rockfall at the mouth of Nankoweap Creek dammed the river, flooding Stanton's Cave.

The mouth of Nankoweap Creek (Figure 9-1a) is approximately 32 kilometers downstream from Stanton's Cave. Marble Canyon widens in the Nankoweap area (Figure 9-1b) and several Quaternary deposits have accumulated in the broad valley. These consist of alluvial deposits derived from Nankoweap Creek, Little Nankoweap Creek, and the Colorado River, as well as colluvial debris derived from the Canyon walls.

Previous investigations of the Nankoweap area by Hamblin and Rigby (1968), Péwé (1968), and the geologic map of the Grand Canyon (Huntoon et al. 1980) indicate that alluvial deposits are present at the mouth of Nankoweap Creek. Péwé (1968:44) noted the distinctive appearance of the rubbly material that forms the ridges on either side of the creek and suggested that the deposits were remnants of an alluvial fan or dissected mudflow. In an overview of mass wastage and rock movement in the Grand Canyon, the colluvial deposits in the Nankoweap area were not discussed (Ford et al. 1974). Work by Hereford (1978) pointed out that the composition of the rubbly material indicated that it was derived from the Paleozoic sedimentary rocks near the top of the almost vertical east wall of Marble Canyon and suggested that it was emplaced by a rockfall.

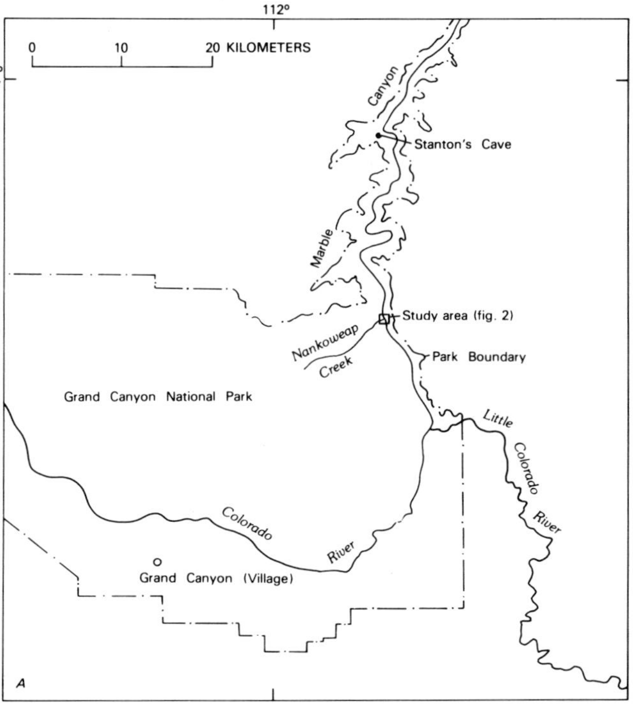

Figure 9-1A. Marble Canyon and Stanton's Cave in the eastern Grand Canyon, Arizona.

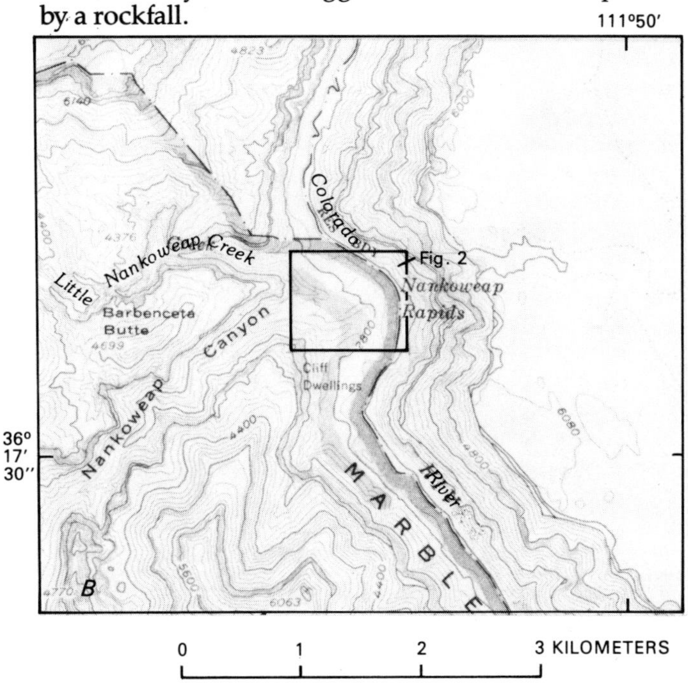

Figure 9-1B. Topographic map of the Nankoweap area from the Nankoweap 15′ quadrangle (1954), contour interval 80 feet.

Reconnaissance studies were done in May 1976, and detailed studies were conducted in November 1976. At that time, a geologic map was made at an approximate scale of 1:7000; elevations were determined by plane table and alidade. Several persons, Robert C. Euler, Thor N.V. Karlstrom, and J.P. Schafer have helped clarify my understanding of the Nankoweap area. I was helped in the field by James A. Whitfield.

Possible Mechanisms for Flooding Stanton's Cave

Two mechanisms, unrelated to events in the Nankoweap area, could have flooded Stanton's Cave. At Lava Falls, 204 kilometers below the Nankoweap area, the river was dammed by lava flows to a maximum elevation of approximately 945 meters (Hamblin 1974), which is 18 meters above the mouth of the cave. The oldest of these flows is dated at 1.2 m.y. This dam provides a remote but nonetheless plausible mechanism for inundating the cave. It is not certain, however, that the mouth of the cave was exposed by that early time because Marble Canyon may not have incised to the elevation of the cave, or the channel may not have been wide enough to expose the entrance to the cave. The river was also dammed by landslides in the Surprise Valley area (Huntoon 1975), 136 kilometers below Nankoweap Creek. At present it is not known when these slides occurred or if an extensive lake formed behind them, consequently it is difficult to associate them with flooding of Stanton's Cave.

The second mechanism involves exceptionally large floods. Assuming that the late Pleistocene-channel geometry was comparable to its present form, Cooley (1971, personal communication to R.C. Euler) estimated that a discharge of approximately 283,000 m^3/s would overtop the mouth of the cave. This discharge is 33 times greater than the largest historic flood of the Colorado River; indeed, it is comparable with catastrophic Pleistocene floods that dissected the channelled scablands of eastern Washington, which had discharges greater than 500,000 m^3/s. These floods are probably the largest discharges of fresh water documented in the geologic record (Baker 1973). A flood of such magnitude on the Colorado River would be rare, and it is difficult to calculate the recurrence interval of such a rare event from a short historic record (Kockel and Baker 1982). If a flood of this magnitude occurred, it probably did not originate from rainfall or snowmelt, but rather from the catastrophic draining of a large possibly ice-dammed lake. Evidence of Pleistocene lacustrine deposits or the existence of a large lake has not been found in the Colorado River drainage basin (Montagne 1972). While it is not possible to rule out flooding of Stanton's Cave by the catastrophic release of ponded water, it seems unlikely because no evidence exists of widespread impoundment of the Colorado River above the cave.

It is impossible to state positively that Stanton's Cave was not inundated by an extremely large flood or by the lake that formed behind the dam at Lava Falls. However, these methods of flooding the cave seem unlikely because the conditions necessary for a flood of the required magnitude did not exist and because of the great distance and age of the dam at Lava Falls. In the following discussion of the Pleistocene geology of the Nankoweap area, it is inferred that the river was temporarily blocked at the mouth of Nankoweap Creek during the middle to late Pleistocene and that the obstruction, which probably originated as a rockfall from the east wall of Marble Canyon, was near the elevation of the cave.

Surficial Geology of the Nankoweap Area

The surficial geology of the study area is shown in Figure 9-2. The youngest deposits consist of interbedded silt, sand, and gravel deposited by the Colorado River and Nankoweap Creek. The youngest unit consists of a floodplain and low terrace; the oldest unit forms a terrace as much as 8 meters above the Colorado River and 2 to 4 meters above the young Holocene alluvial unit. These deposits are essentially unweathered and lie within or near the historic flood levels of the Colorado River. They are considered to be entirely of Holocene age; consequently they are younger than the driftwood and postdate the flooding of Stanton's Cave.

Alluvium and colluvium are the predominant types of Pleistocene deposits in the Nankoweap area. The oldest alluvial deposit, which crops out on either side of Nankoweap Creek, is a gravel consisting of rounded

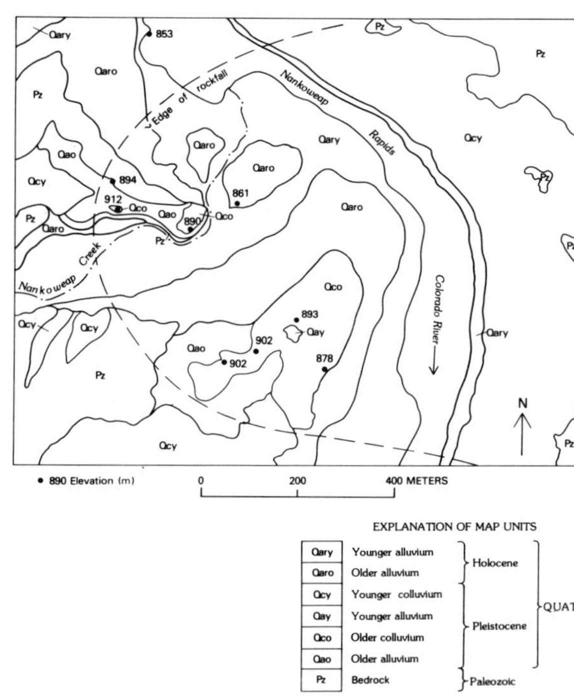

Figure 9-2. Surficial geologic map of the Nankoweap Rapids area.

pebbles, cobbles, and boulders derived from the Paleozoic bedrock in Nankoweap Creek as well as clasts of quartzite and porphyry transported by the Colorado River. The abundance of sediment derived from Nankoweap Creek suggests that this unit was an alluvial fan built into Marble Canyon 10 to 25 meters above the present elevation of Nankoweap Creek and the Colorado River. The height of the deposit above the river and its extensive dissection by Nankoweap Creek suggest that it is Pleistocene in age.

A distinctive deposit with a poorly sorted, rubbly texture (unit Qco, Figure 9-2) overlies the older alluvium. The measured thickness varies from a minimum of 2 meters to a maximum of between 41 and 57 meters. Thickness is variable because the base of the deposit is uneven. The deposit is composed mainly of large angular blocks of Permian Kaibab Limestone as much as 4 meters on a side. Smaller blocks, in order of decreasing abundance, are sandstones of the Supai Group (Pennsylvanian and Permian) and Coconino Sandstone (Permian). The clasts of Kaibab Limestone are weathered; commonly the surface is etched to a depth of 1 to 10 cm. A few clasts are more deeply weathered and show cavernous weathering.

The matrix generally forms less than 20 percent of the deposit and consists mainly of angular clasts of very fine to very coarse sand. Silt and clay are present but form a minor portion of the matrix. Rounded pebbles of quartzite and porphyry similar to those carried by the Colorado River, but not by Nankoweap Creek, are widely scattered throughout the deposit (Figure 9-4). The deposit is massive (Figure 9-3), bedding of any scale or type is absent, suggesting that it was deposited during a single event.

On the south side of Nankoweap Creek, the colluvium is overlain by a lag deposit (unit Qay, Figure 9-2) less than 50 cm thickness composed of rounded quartzite pebbles similar to those in the channel of the Colorado River. This gravel and a large water-worn clast of Kaibab Limestone near elevation 878 meters (Figure 9-2) indicate that the Colorado River flowed over the rubble. This suggests that the rubble filled the channel from wall to wall, thereby damming the river. The lag deposit probably records the dissection and removal of the dam.

The older colluvium probably originated from the east wall of Marble Canyon. This is suggested by the composition of the deposit; specifically the abundant clasts of Kaibab Limestone, Coconino Sandstone, and sandstones of the Supai Group as well as the pebbles of quartzite and porphyry. A nearby source of the Paleozoic clasts is the upper part of the east wall of Marble Canyon. Parts of the east wall are vertical and the material would be deposited by sliding and falling, as in a rockfall. The pebbles were probably swept from the channel as the rockfall passed over the Colorado River.

An alternative source area is the west side of Marble Canyon, but this is unlikely for several reasons. If derived from the west, the rubble would not contain clasts from the Colorado River because such material does not occur in Nankoweap Creek. Moreover, to transport the large blocks of bedrock along the comparatively low gradients of Nankoweap Creek (Figure 9-1b) would probably require some mixture of clay and water such as in a debris flow, but clay forms an insignificant portion of the colluvium. Finally, debris flows require a ready source of rubble such as a thick, potentially unstable accumulation of talus (Statham 1977:93). In a study of Pleistocene debris flows along Comanche Creek in the eastern Grand Canyon, Schafer (1976, personal communication) found that they originated from thick accumulations of rubble. Thick rubble deposits do not occur in the lower reaches of Nankoweap or Little Nankoweap creeks (Huntoon et al. 1980). The closest rubble deposits are more than 4 kilometers from the Nankoweap area, and no evidence exists that material has flowed from them down Nankoweap or Little Nankoweap creeks.

Figure 9-3. Exposure of the older Pleistocene colluvium near elevation 890 m (Figure 9-2). Base of outcrop is about 10 m in length.

Figure 9-4. Small clasts and matrix of the older colluvium. Note rounded quartzite pebbles probably of Colorado River origin above scale. Scale is 15 cm in length.

It seems reasonable to conclude that the older colluvium originated from the east side of Marble Canyon. Figure 9-5 shows the position of the colluvium in Marble Canyon and its relationship to the other Pleistocene deposits. The clasts of Kaibab Limestone traveled vertically somewhat more than they moved horizontally, and they must have fallen freely through space as they passed over the Redwall Limestone which forms a vertical wall on the east wall of Marble Canyon. The event, therefore, was a rockfall rather than a rockslide or avalanche. The rockfall was probably initiated by basal undercutting of the upper cliff by erosion of the relatively soft Hermit Shale (Permian).

On the north and south side of Nankoweap Creek, the older colluvium and alluvium are overlain by talus (unit Qcy Figure 9-2). The deposit is dominated by blocks of Muav Limestone (Cambrian) as large as 1 to 2 meters on a side. The blocks were deposited beneath steep slopes and cliffs on the gentle slopes formed on the Bright Angel Shale (Cambrian). The deposit is only moderately dissected, slightly weathered, and is overlapped by Holocene alluvium, suggesting that it is of late Pleistocene age.

The talus deposits are cemented with travertine and occur throughout Marble Canyon from Nankoweap Creek to the Little Colorado River (Péwé 1968, Figure 10). The extensive travertine cementation resulted from numerous springs, whose water was charged with $CaCo_3$, percolating through the talus. The numerous springs present during cementation of the talus imply that the climate was colder and probably wetter than at present. Paleoclimatic studies by Cole (1982) in the eastern Grand Canyon, suggest that during the last full-glacial period, 15,000 to 21,000 years ago, the climate was colder in all seasons and probably wetter than it is today. It is likely that the travertine-cemented talus formed during this time of reduced temperature and increased moisture when springs were more numerous than at present.

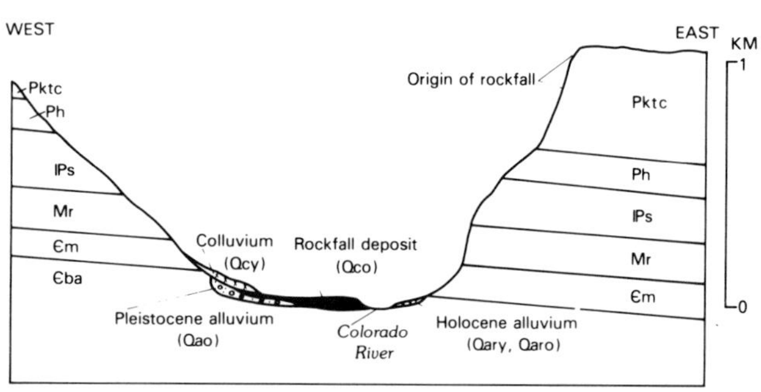

Figure 9-5. Generalized cross section of Marble Canyon at Nankoweap Rapids showing the rockfall and other surficial deposits. Vertical and horizontal scale equal. Єba = Bright Angel Shale; Єm = Muav Limestone; Mr = Redwall Limestone; Ps = Supai Group; Ph = Hermit Shale; Pk, t, c = Kaibab Limestone, Toroweap Formation, and Coconino Sandstone undifferentiated; Qao = older Pleistocene alluvium; Qco = older Pleistocene colluvium; Qaro = older Holocene alluvium; Qary = young Holocene alluvium.

Of the several Pleistocene deposits in the Nankoweap area, the older Pleistocene colluvium is the only one capable of blocking the Colorado River. The thickness and reconstructed outline of the deposit, which is suggested by the preserved shape of the debris and by assuming that it originated from the east side of the Canyon, imply that it was probably large enough to obstruct the river (see approximate edge of rockfall, Figure 9-2). The elevation of the surface of the deposit ranges from 890 to 912 meters, which is between 15 and 37 meters below the mouth of Stanton's Cave. Although the differences between the altitude of the deposit and the mouth of the cave seem substantial, one should recall that they are maximum values. Field evidence indicates that the rockfall has been extensively eroded by the Colorado River and Nankoweap Creek. It is possible, therefore, that the top of the rockfall was as least as high as the mouth of Stanton's Cave and that the cave may have been flooded shortly after the rockfall occurred.

Age of the Rockfall

The absolute age of the rockfall is not known, but field relationships suggest that it probably occurred during the middle to late Pleistocene. A basalt flow in the western Grand Canyon, 204 kilometers down from the Nankoweap area, is situated in the channel near river level and is dated at 1.2 m.y. (Hamblin 1974). It is possible that Marble Canyon was also near its present depth at that time, suggesting that the older alluvium, which occurs in a position above the river similar to the basalt flows, is no older than the 1.2 m.y. flows. The rockfall, therefore, must be younger than 1.2 m.y. because this date is the maximum age of the older alluvium. The older alluvium is at the same height above the river as the youngest (or lowest) Quaternary gravel deposit mapped by Phoenix (1963) and Bush (1983) near the head of Marble Canyon at Lees Ferry and along the Paria River; consequently, they are probably the same relative age. The absolute age of the gravel at Lees Ferry and the Paria River is unknown, but gravel deposits in the nearby Little Colorado River valley at a similar position are between 150,000 and 500,000 years old (Ulrich et al., in press). It is likely, therefore, that the maximum age of the rockfall is between 150,000 and 500,000 years. The minimum age of the rockfall must be greater than the age of the younger colluvium, which probably formed during the last full-glacial period 15,000 to 21,000 years ago.

Conclusions

The composition and texture of a dissected middle

to late Pleistocene deposit on the west side of the Colorado River near Nankoweap Rapids suggests that it originated as a rockfall from the east wall of Marble Canyon. The reconstructed area of the deposit indicates that it was large enough to block the river. This interpretation is supported by a thin deposit of river gravel at the top of rockfall that indicates that the river flowed over the debris. The present elevation of the rockfall, which is lower than its original height because of extensive dissection, is close to the elevation of Stanton's Cave. Pleistocene driftwood in Stanton's Cave may have floated into the cave on the lake that formed behind the rockfall dam.

An alternative explanation is that the cave was flooded by a lake that formed after lava flows dammed the river 1.2 m.y. ago at Lava Falls, 204 kilometers below the Nankoweap area. Though plausible, this mechanism is complicated by the remoteness of the dam and possibility that the mouth of the cave was not exposed 1.2 m.y. ago. Another explanation is that the cave was flooded during exceptionally high runoff from snowmelt or rainfall, or from catastrophic draining of a large, possibly ice-dammed, lake. The discharge necessary to flood the cave has been estimated to be 283,000 m^3/s. This discharge is so large that its recurrence interval cannot be accurately esitmated, and it is within an order of magnitude of the largest flood that resulted from the catastrophic draining of Pleistocene Lake Missoula, which is the largest discharge of fresh water documented in the geologic record. The possibility of such large floods on the Colorado River seems unlikely because of the absence of widespread lacustrine deposits.

Footnotes

1. Not all geologists are in agreement with the hypothesis presented in this chapter. Ongoing studies of gravel deposits in the area, at present in a preliminary stage, suggest that additional hydrological and geomorphological factors affecting past levels of the Colorado River may have been in part responsible for the deposition of the driftwood in Stanton's Cave — Editor.
2. A more recent ^{14}C date, derived since this was written, from a piece of the same log as that dated by the University of Arizona, is 43,700 + 1800 and -1500 years B.P. (U.S.G.S. No. 1774) analyzed at the radiocarbon laboratory of the U.S. Geological Survey at Menlo Park, California, by S.W. Robinson (personal communication, 1984) — Editor.

References

Baker, V.R.
1973 *Paleohydrology and sedimentology of Lake Missoula flooding in eastern Washington.* Geological Society America Special Paper 144, 79 pages.

Bush, A.L.
1983 *Geologic map of the Vermilion Cliffs-Paria Canyon instant study area and adjacent wilderness study areas, Coconino County, Arizona, and Kane County, Arizona.* U.S. Geological Survey, Miscellaneous Field Studies Map, MF-1475-A.

Cole, Kenneth
1982 Late Quaternary zonation of vegetation in the eastern Grand Canyon. *Science* 217:1142-1145.

Euler, R.C.
1978 Archeological and paleobiological studies at Stanton's Cave, Grand Canyon National Park, Arizona — a report of progress. *National Geographic Society Research Reports,* 141-162.

Ford, T.D., G.H. Billingsley, Jr., P.W. Huntoon, and W.J. Breed
1974 "Rock movement and mass wastage in the Grand Canyon" *In* W.M. Breed and E.C. Roat, ed., *Geology of the Grand Canyon.* Flagstaff and Grand Canyon: Museum of Northern Arizona and Grand Canyon Natural History Association, pages 116-128.

Hamblin, W.K.
1974 "Late Cenozoic volcanism in the western Grand Canyon" *In* W.J. Breed and E.C. Roat, ed., *Geology of the Grand Canyon.* Flagstaff and Grand Canyon: Museum of Northern Arizona and Grand Canyon Natural History Association, pages 142-169.

Hamblin, W., and J.K. Rigby
1968 Guidebook to the Colorado River, Part 1: Lee's Ferry to Phantom Ranch in Grand Canyon National Park. Provo, Utah: *Brigham Young University Geology Studies* 15:3-84.

Hereford, Richard
1978 An ancient rockfall at Nankoweap Rapids in Marble Canyon, Grand Canyon National Park, Arizona (abs.). *Geological Society America Abstracts with Programs* 10:109.

Huntoon, P.W.
1975 The Surprise Valley landslide and widening of the Grand Canyon. *Plateau* 48:1-12.

Huntoon, P.W., G.H. Billingsley, Jr., W.J. Breed, J.W. Sears, T.D. Ford, M.D. Clark, R.S. Babcock, E.H. Brown
1980 *Geologic map of the Grand Canyon National Park, Arizona.* Grand Canyon and Flagstaff: Grand Canyon Natural History Association and Museum of Northern Arizona, scale 1:62,500.

Kockel, R.C. and V.R. Baker
1982 Paleoflood hydrology. *Science* 215:353-361.

Montagne, J.M.
1972 "Quaternary system, Wisconsin glaciation" *In* Geologic atlas of the Rocky Mountain region. Denver: *Rocky Mountain Association Geologists,* 257-260.

Péwé, T.L.
1968 *Colorado River guidebook, Lees Ferry to Phantom Ranch.* Private publication, 78 pages.

Phoenix, D.A.
1963 *Geology of the Lees Ferry Area, Coconino County, Arizona.* U.S. Geological Survey Bulletin 1137, 86 pages.

Statham, Ian
1977 *Earth surface sediment transport.* Oxford University Press, 184 pages.

Ulrich, G.E., G.H. Billingsley, Jr., Richard Hereford, E.W. Wolfe, L.D. Nealey, R.L. Sutton
in press *Geology, structure, and uranium deposits of the Flagstaff, 1x2' quadrangle, Arizona.* U.S. Geological Survey Miscellaneous Investigations Map I-1446, scale 1:250,000.

Chapter 10

Polarity of River-Flood Silt
in Stanton's Cave
Marble Canyon, Arizona

by

Donald P. Elston
U.S. Geological Survey
Flagstaff, Arizona

Abstract

In Stanton's Cave, Marble Canyon, Arizona, unconsolidated silt and associated driftwood appear to have accumulated during a single episode of flooding of the Colorado River. Thirty oriented samples of the silt, collected for paleomagnetic analysis, were obtained from four sites in two trenches. The silt was impregnated with lacquer in order to collect coherent cubical samples for the measurement of paleomagnetic directions. Stepwise demagnetization analysis in alternating fields revealed only stable, normal polarity directions, most of which cluster closely about the direction of the present magnetic field. No second components of magnetization were observed. The samples carry a normal polarity remanence of apparent depositional origin, presumably residing in finely disseminated detrital magnetite. Because of this, the silt and associated driftwood logs in Stanton's Cave accumulated no earlier than about 720,000 years, the nominal age of the lower boundary of the Brunhes normal polarity epoch. Because of the lack of any postdepositional cementation, secondary alteration, and secondary magnetization, and in view of a generally modern aspect of the associated driftwood, the flood-related deposit is inferred to be much younger than the maximum age permitted by the paleomagnetic data. Since the preparation of this report, this inference has been verified by a radiocarbon age of approximately 43,700 years for a sample of the driftwood that previously had been reported to be more than 35,000 years old.

Introduction

Paleomagnetic results presented here are derived from oriented samples collected from Stanton's Cave, Marble Canyon, Arizona, May 5 and 6, 1982. The study was undertaken to check the validity of a reverse polarity direction that had been reported from a few of a small group of samples of silt collected in 1976 (Euler 1978:158), and also to obtain a data set adequate for evaluating the polarity of the entire deposit. Reverse polarity would indicate an age of deposition greater than about 720,000 years, the boundary between the Brunhes normal polarity epoch and the older Matuyama reverse polarity epoch (for example, see Lowrie and Alvarez 1981). Such an age would be much greater than a minimum age of about 35,000+ years or 43,700 + and - years that had been reported from the radiocarbon dating of driftwood associated with the deposit (see discussion elsewhere in this volume). For this reason, it was necessary not only to collect from the interval from which the alleged reverse polarity samples had come, but also to collect from other suitable parts of the deposit exposed in the trench walls.

Stratigraphy

The stratigraphy of the deposit is depicted in two trench profiles developed by R.C. Euler (see Figure 2-5, this volume). Parts of these profiles are shown in Figure 10-1. Also shown on this figure is the approximate distribution of the individual samples collected at the four sites. One site is in the North-South Trench, and three sites are in the East-West Trench. Outlines of the stratigraphic units of Euler are shown on Figure 10-1. For an explanation of the units, see the original diagrams elsewhere in this volume.

Samples at the four sites were obtained from the finest grained materials exposed in the trenches, primarily silt, which accumulated at the time that flood waters and driftwood entered the cave; upon receding, these waters left behind a heterogeneous deposit consisting of a tangled mass of driftwood, the spaces within which are filled with silt and sand. Clastic materials exposed in the trench walls away from the four samples sites are mostly coarser than silt, and because of this were considered much less likely to contain a reliable paleomagnetic record of depositional origin.

Sampling: The silt is poorly stratified at best, and lacks interstitial cement. It is found in ill-defined layers and as small pockets within a complex mass of driftwood. Thus, to collect oriented samples, a bonding agent was needed that would impregnate deeply, dry quickly, and would not be so liquid that it would disturb the friable silt. Waterglass (sodium silicate) penetrated poorly and dried slowly; however, clear lacquer, sprayed repeatedly as a mist, penetrated the silt without disturbing it and cemented it deeply enough to allow cubical samples to be carved in the exposures. The face of each sample carved in the trench wall was individually oriented. Plastic boxes (1x1¾'') then were placed over the cubes and the samples detached from the outcrop. This process allowed 30 samples to be painstakingly collected during a 1½ day span of time. Samples from Grid 0-19 were collected by Joseph Rosenbaum, samples from Grid H-H were collected by Charles Naeser, samples from Grid F-F were collected by Hugh Rieck, and samples from Grid A-A were collected by James Scott and Gary Olhoeft.

Paleomagnetic Directions

All samples exhibit normal polarity, with most directions (declinations and inclinations) clustering about the direction of the present magnetic field (Figure 10-2). Directions of natural remanent magnetization (NRM) are more tightly grouped than directions following partial demagnetization in an alternating field. In Figure 10-2, compare NRM directions and directions following cleaning in a 100 oersted (oe) alternating field (NRM, n=22, Dec. 353.9°, Inc. 58.4°, alpha 95°=4.7°, k=45.5°; 100 oe, n=22, Dec. 354.7°, Inc. 56.0°, alpha 95°=6.1°, k=27.1°). In spite of increased scatter with "cleaning" the directions from the four sites and the overall mean direction remained well grouped up through a cleaning step of 300 oe.

One of the sample groups, F-F, displays an elongate "streak" that reflects departures in both declination and inclination for some of the samples of this group. The streaking may have arisen from a slightly greater coarseness of the material at this site, as well as from difficulties encountered in sampling the very small pockets of fine-grained material. Nonetheless, some of the samples from this site plot in the main group of directions derived from the other three sites.

Progressive stepwise demagnetization analysis was carried out in the range of 25 to 950 oe on at least two samples from each of the four sample groups. This procedure allowed the stability of the magnetization to be assessed, and the existence of possible multicomponent magnetizations, reflecting alterations and secondary magnetizations, to be recognized. Good stability was found, and no secondary components of magnetization were observed. Representative equal-area and orthogonal plots are shown in Figures 10-3 and 10-4, respectively.

During progressive partial demagnetization analysis, significant scattering in direction and departures in linearity of the orthogonal plots were found to occur mainly above the 500 oe cleaning steps. Small shifts, marked by a clustering of directions on the equal-area plots and by linear traces on the orthogonal diagrams that trend toward the origins, are seen in mainly in the 100 to 300 oe demagnetization steps. The mean direction for the 100 oe cleaning step does not differ significantly from the mean NRM direction. However, the grouping of directions of the individual samples at this cleaning step is poorer than the grouping of NRM directions, reflected by a larger cone at the 95 percent confidence level and a smaller value for the precision parameter (kappa). (Compare the mean directions and statistics for 22 samples given earlier.) Therefore, the progressive demagnetization data are interpreted to reflect the removal of only a single component of magnetization, one acquired at the time of deposition. Because stability is lost in degmagnetizing fields above 500 oe, the magnetization is considered to reside in finely disseminated magnetite. Hematite does not respond appreciably in demagnetizing fields below 1000 oe, and only partially in fields of 2000 oe and higher.

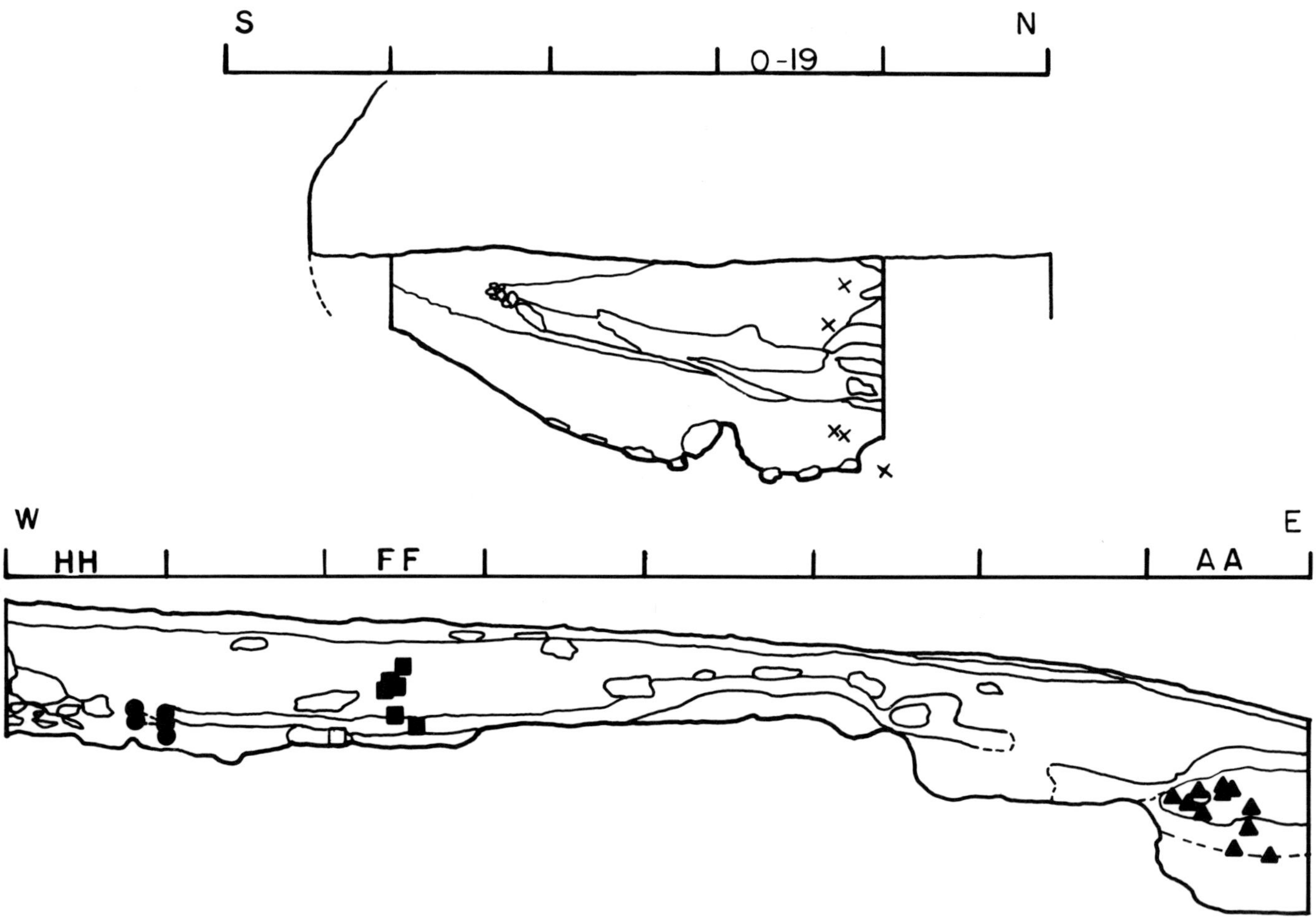

Figure 10-1. Generalized sections of trench walls in Stanton's Cave showing approximate locations of samples from four sample groups. One site is in the 0-19 block of the North-South Trench, and three are in parts of the East-West Trench labeled H-H, F-F, and A-A. Not all sample locations are shown because some overlapping sample positions lie normal to the plane of the section. For lithology and scale, see original diagrams of Euler, this volume.

Only a slight scattering of directions occurs as a consequence of partial demagnetization in alternating fields of 200 oe and less. This is in contrast to behavior that is normally encountered in lithified deposits. In consolidated and partially lithified deposits, an improved clustering of paleomagnetic directions commonly is observed with the removal of "soft" secondary components of magnetization in low alternating fields. In such cases, the improved groupings are then considered to more closely approximate the direction of the ambient field at the time of accumulation than the NRM directions.

It is rare that completely unconsolidated sediments such as those found in Stanton's Cave are investigated for their paleomagnetic record — sediments that have undergone no apparent post-depositional compaction and cementation, and no apparent alteration. Such preservation appears attributable to the existence of a dry environment since the time of accumulation, support for which is seen in the complete lack of secondary components of magnetization in any of the samples, all of which were subjected to progressive demagnetization analysis to 150 oe. No alteration attributable to subsequent prolonged wetting or the percolation of waters appears to have taken place following the episode of flooding. Thus, the flood-related deposit in Stanton's Cave appears to be very young on paleomagnetic as well as geologic grounds.

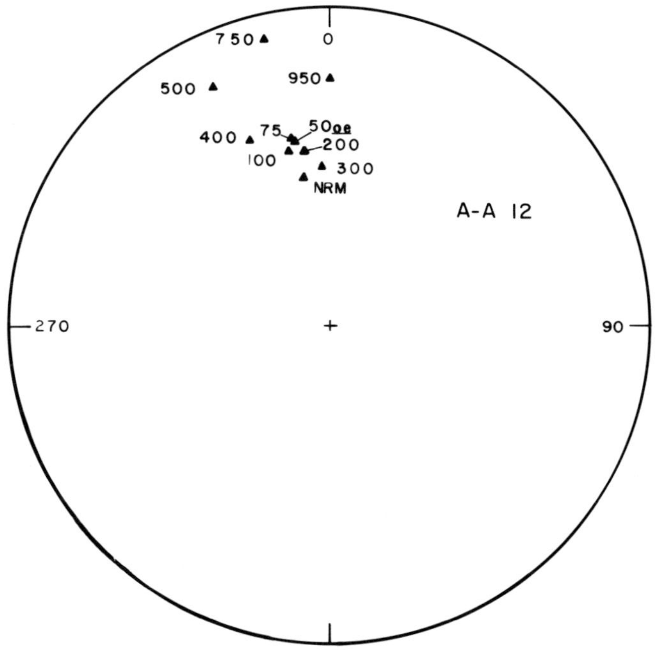

Figure 10-3. Equal area plot showing directions resulting from progressive alternating field (AF) demagnetization in the range of 50-950 oersteds.

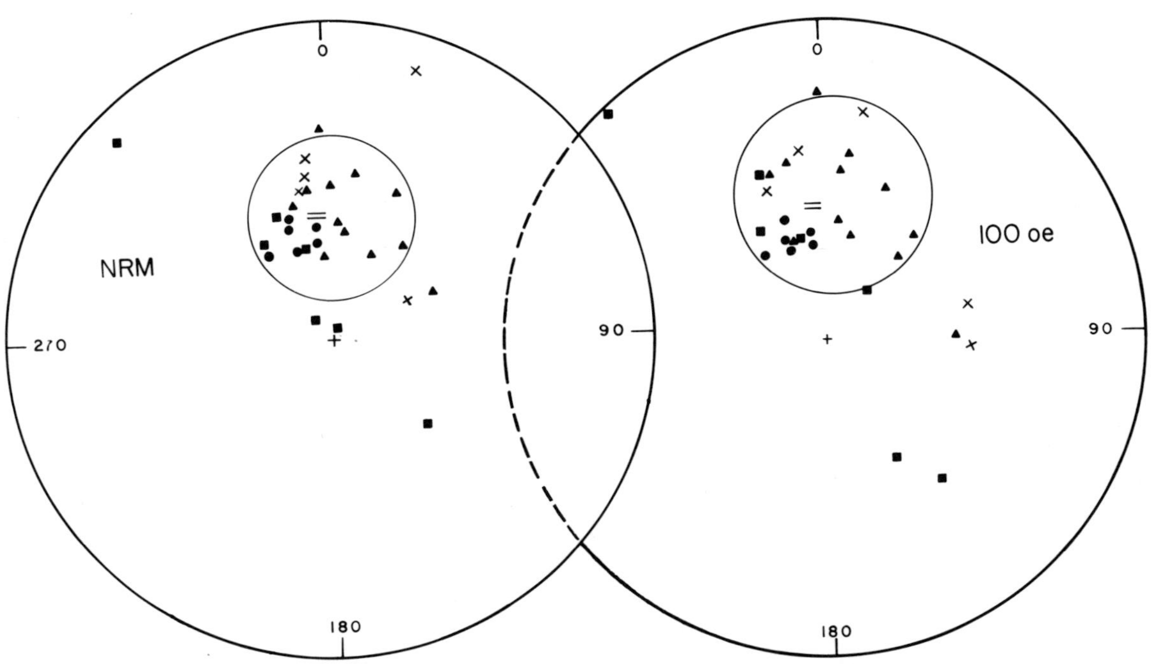

Figure 10-2. Equal area plots of declination and inclination of 30 oriented samples of silt from Stanton's Cave. Solid symbols denote positive inclinations, which plot on lower hemisphere. Symbols (crosses, dots, squares, and triangles relate to sites shown on Figure 10-1. Circles enclose the 22 samples used for the calculation of a mean normal polarity direction. Equal signs denote mean directions, which are given in the text. Note the increase in scatter of the individual directions with partial demagnetization in an alternating field of 100 oersteds (oe), but that the mean direction is virtually unchanged.

Although plots of directions for all 30 samples are shown on Figure 10-2, only 22 samples were used for the calculation of a mean direction. Eight samples that lie outside of the main cluster of directions were discarded. The resulting mean NRM direction is considered to be best estimate of the ambient field direction at the time of deposition.

Summary

No evidence exists for reverse polarity in any samples collected for this study of silt from Stanton's Cave. The mean direction obtained before cleaning (the NRM direction) provides the best estimate of the ambient field direction when the Colorado River deposited silt and associated driftwood in the cave. Because the driftwood is identical to modern species, and because the sediment lacks post-depositional alterations and secondary magnetizations, the deposit is certainly less than 720,000 years old and probably is very much younger. Subsequent to the preparation of this manuscript, the foregoing inference was verified by a radiocarbon age from a sample of driftwood that had previously been reported to be older than 35,000 years. The sample, USGS 1774, analyzed at the radiocarbon laboratory of the U.S. Geological Survey at Menlo Park, California, by S.W. Robinson (personal communication, 1984), has a finite age of 43,700 +1800 & -1500 years B.P.

Acknowledgments

I thank D. Champion and R. Hereford for their reviews of this manuscript.

References

Euler, Robert C.
1978 Archeological and Paleobiological Studies at Stanton's Cave, Grand Canyon National Park, Arizona — A Report of Progress. *National Geographic Society Research Reports, 1969 Projects* 141-162.

Lowrie, William and Walter Alvarez
1981 One hundred million years of geomagnetic polarity history. *Geology* 9:392-397.

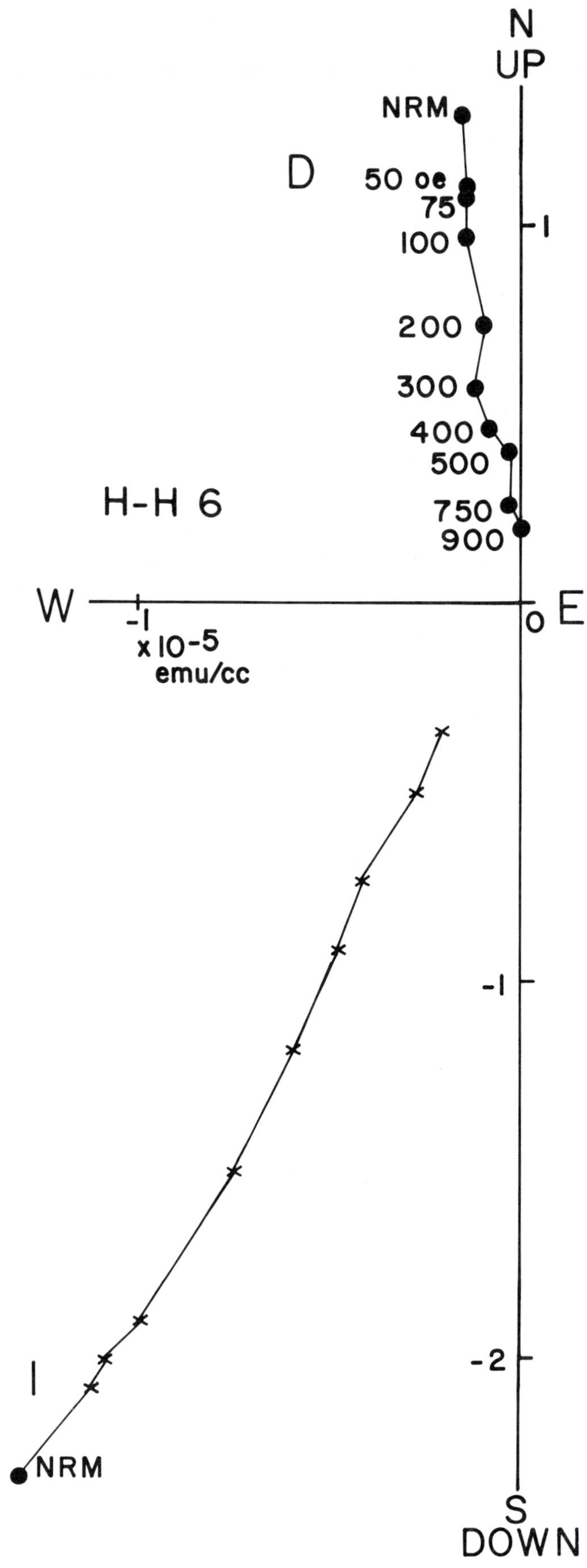

Figure 10-4. Orthogonal demagnetization diagram showing projection of vector end point during progressive alternating field demagnetization analysis. Closed circles, horizontal plane; Xs, vertical plane.

Chapter 11

Paleoecology of Stanton's Cave Grand Canyon, Arizona

by

Eleanora I. Robbins*
U.S. Geological Survey
Reston, Virginia

and

Paul S. Martin and Austin Long
Department of Geosciences
University of Arizona, Tucson

Publication No. 746, Department of Geosciences,
University of Arizona

*Formerly Eleanora R. Iberall

Abstract

Fecal pellets and fossil bones are evidence that both mountain sheep *(Ovis canadensis)* and another artiodactyl, probably an extinct mountain goat *(Oreamnos harringtoni)*, occupied Stanton's Cave between the late Pleistocene and the Holocene. Plant fragments in the fecal pellets showed that the animals foraged on xeric plants similar to those still found in the vicinity of the cave. A shift from large to small pellets about 10,800 years ago represents the end of occupation of the cave by *Oreamnos.* Major climatic change during the time representing habitation by both animals in the Grand Canyon was more apparent in pollen in the cave earth than in plant epidermis in the pellets.

Introduction

Stanton's Cave is at kilometer 51 (51 km from Lees Ferry, Arizona, measured along the Colorado River). It is on the north rim of Marble Canyon in Arizona at 36°30'N, 111°50'W (Figures 11-1 and 11-2). The cave opens at 44 meters above the present level of the Colorado River at an elevation of 927 meters. It penetrates 122 meters into the Redwall Limestone of Mississipian age. Below Stanton's Cave and immediately downstream is a lush green environment created by a connected series of springs. The springs were named Vaseys Paradise by John Wesley Powell, the explorer-geologist, on his first trip through the Grand Canyon in 1869.

Several plant communities are established in the vicinity of the cave and represent southern exposure associations of Great Basin desert plants. A partial listing of these plants is included in Table 11-1 (and see Hevly, this volume). Desert-scrub and riparian communities, typical of the inner gorge of the Colorado River, dominate at water level (Hoffmeister 1971). Nonnative plants are joining the riparian community since the completion of the Glen Canyon Dam in 1963 (King and Sigleo 1973). Before 1962 the high-water mark of seasonal flooding of the Colorado determined the lowest point along the river that perennial species could grow. On the talus slopes in front of the cave, three species of grasses and 49 species of herbaceous and woody plants have been identified in traverses from the Colorado River to the cave. Common woody plants include *Baccharis sergiloides, Ephedra* spp., *Fallugia paradoxa,* and *Yucca angustissima.* Herbaceous vegetation is primarily grasses, composites, and legumes. The springs at Vaseys Paradise primarily have a dense cover of phreatophytes, including 58 species of grasses, sedges, rushes, orchids, ferns, and scrophularias. *Cercis occidentalis, Juniperus osteosperma, Salix* sp., and *Tamarix pentandra* grow in the vicinity of the springs. Up the steep slopes of the Grand Canyon, *Ephedra, Agave,* and *Opuntia,* as well as clumps of grasses, cling to cracks and gain holdfast along ledges. Along South Canyon above the cave, and on the Grand Canyon's Kaibab Plateau rim, five species of grasses and 28 species of predominantly low woody shrubs have been identified. Isolated clumps of *Yucca* are part of an open association of annual grasses and weed species alternating with patches of bare ground. Still higher, between 1620 and 1890 meters, Kane Canyon (which contains springs) is dominated by a piñon-juniper community containing 54 species of herbs and woody shrubs.

In general, because pollen production of desert plant communities is small, tree species such as *Pinus, Juniperus,* and *Quercus,* as well as *Ephedra,* that are wind pollinated and that produce abundant pollen, typically dominate desert pollen profiles. The modern pollen rain outside Stanton's Cave is comprised of more than 20% each *Pinus* and *Juniperus;* 10-15% each Chenopodiaceae-*Amaranthus* (Cheno-Ams), and low-spine Compositae; 5% each *Quercus, Ephedra nevadensis-viridis*-type and Gramineae; and trace amounts to 1% each *Abies* and *Alnus* (King and Sigleo 1973). The surface layer of

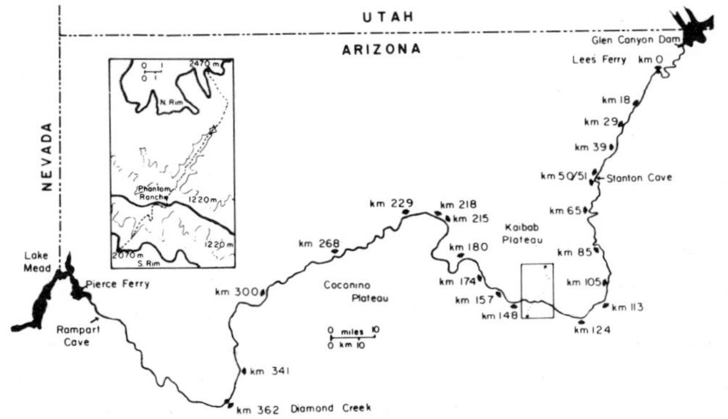

Figure 11-1. Location of Stanton's and Rampart caves, Grand Canyon, Arizona.

Figure 11-2. Stanton's Cave and Vaseys Paradise. Vaseys issues as three springs along bedding planes in the Redwall Limestone. The picture shows the relative abundance of vegetation. Note the pre-Glen Canyon dam floodline. Paul S. Martin, photographer.

the cave itself contains: 50% *Pinus;* 10-20% each *Ephedra, Juniperus,* Cheno-Ams, and Compositae; 8% *Artemisia;* and trace amounts of *Quercus, Salix, Sarcobatus,* and Graminae pollen.

The animal community inside Stanton's Cave consists primarily of small mammals such as pack rats *(Neotoma)* and bats (species unknown). Bighorn sheep *(Ovis canadensis)* have been seen on the Grand Canyon rims, where the closest sighting to the cave was near Redwall Cavern less than 2 kilometers downstream. Two sheep were observed there on August 8, 1970 (Grand Canyon National Park records).

In historical times, before the Glen Canyon Dam, the river was at flood maximum in early summer and was otherwise low for months at a time. Now the Colorado River provides a continuous water supply, and Vaseys Paradise is typical of the permanent springs issuing from the side canyons and supplementing the river flow.

Excavation in Stanton's Cave revealed bones of bighorn sheep *(Ovis)*, extinct mountain goat *(Oreamnos)*, camel *(Camelops)*, horse *(Equus)*, and possibly two other artiodactyls *(Bison*? and *Odocoileus*?). In 1970, Robert C. Euler's party collected all material of interest from a 1 m^2 section (Grid I-I) from the cave surface to the bedrock floor. A 5 cm vertical sampling interval was established, and all cave earth from the grid was sifted through 64 mm (¼ inch) window screen. Along the wall of Grid I-I, pollen samples were taken at 5 cm intervals. The most common fossils removed from the section were lozenge-shaped fecal pellets of artiodactyls (Figure 11-3). Owing to the dry environment typical of caves in the arid Southwest, these were well preserved. Generally, pellets outside caves disintegrate

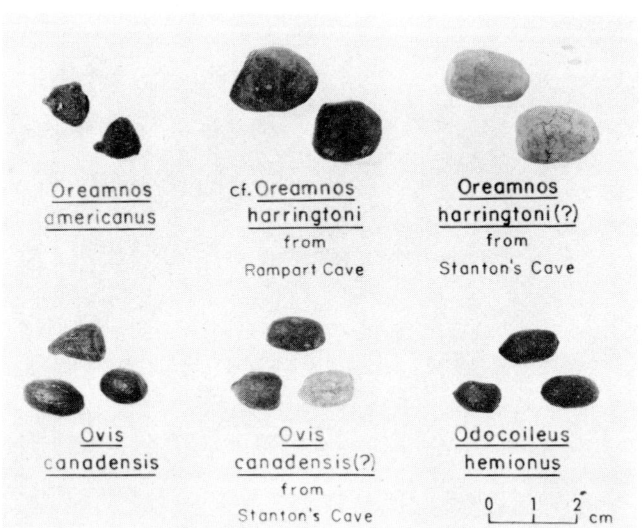

Figure 11-3. Modern and fossil fecal pellets. Note excellent preservation of fossil specimens. *(Oreamnos americanus* — mountain goat; *O. harringtoni* — Harrington's mountain goat; *Ovis canadensis* — bighorn sheep; *Odocoileus hemionus* — mule deer.)

a week or two after the first rain (Welles and Welles 1961).

These fecal pellets provide a unique source of information. In other caves, the presence of modern herbivores, the remains and pellets of extinct herbivores, the content of pack rat nests, and sloth dung have been used to determine factors related to animal and plant community extinctions and climatic change. For example, Laudermilk and Munz (1938) and Long et al. (1974) studied plant matter in the dung of extinct Shasta ground sloths from Rampart Cave and concluded that the sloths ate plants similar to those growing today in the same place near Rampart Cave (Figure 11-1), which is at kilometer 442 downstream from Lees Ferry. In a stratigraphic section of the same sloth dung, Long and Martin (1974) found evidence for periodic occupation of the cave by these animals. Wilson (1942) identified an extensive Pleistocene fauna in Rampart Cave and concluded that several of the species were adapted to colder climatic conditions than are species presently in the area. Pollen from the fill of Rampart Cave (Martin et al. 1961) gave evidence for three climatic shifts, from warm/dry to cool/moist to warm/dry, within the past 35,000 years. On the basis of the contents of fossil pack rat nests, Wells (1966), Wells and Berger (1967), Van Devender and King (1971), Wells and Jorgenson (1964), and Madsen (1973) discussed vegetation changes that led to the present vegetation of the southwestern deserts. Madsen (1973) concluded that through most of late Wisconsin time, vegetation zones were as much as 1500 m lower than they are today in the Great Basin.

Analysis and identification of plant cuticles and epidermal cells to the species level is possible (Stace 1965). Analyses of plant species in fecal pellets of living animals have been instrumental in determinations of their diets (Storr 1961; Hansen 1971; Stewart and Stewart 1970). In this paper, and in Iberall (1972), Robbins analyzes the fecal pellets from Stanton's Cave in order to identify the artiodactyl species responsible for the dung and to identify the plant species they ate.

Preparation of Reference Slides

A reference collection of plant cuticles and epidermal cells was prepared from leaves of 189 species of plants collected fresh in the field or taken from herbarium mounts of other collectors. The plants were collected near Stanton's Cave, starting at the Colorado River (804 m) and ending at Jacob Lake, Arizona (1890 m). Reference slides were prepared by two techniques popular in the current literature — that of Storr (1961), which involves boiling plants in nitric and chromic acid, and that of Sparks and Malachek (1968), which involves grinding plants to particles of a uniform size and then processing them through Hertwig's and Hoyer's solutions. Either process is used to duplicate digestion of plants through a herbivore's gut (Martin 1954; Croker 1959). Identifications were made by comparing fragments of plants found in the fecal pellets to those on the reference slides.

Analysis of Fecal Pellets

In Grid I-I, 4664 pellets were sorted from the cave earth. These were individually weighed and measured. Figure 11-3 shows some of these pellets. Table 11-2 shows the total number of pellets removed from the test pit at each depth. A sharp increase in the number of pellets in any one level, such as the 20-25 cm depth interval, is not considered significant. One bighorn sheep can deposit as many as 500 pellets a day (Welles and Welles 1961).

The measurement of the pellets showed that two different sizes were present. Figure 11-4 shows the weights of pellets found in each 5 cm depth interval of Grid I-I and the weights of pellets known to be from one extinct and three modern animals. Above 20 cm depth, pellets average 0.22 gm and do not exceed 0.6 gm. Below 25 cm depth pellets average 0.45 gm. The relative amount of broken pellets (shown by a stipple pattern in Figure 11-4) increases with depth. In the 20-25 cm depth interval, a bimodal distribution of weights of pellets suggests the presence of two different species of animals. Figure 11-5 shows lengths, widths, and weights of the two apparent pellet groups (above 20 cm and below 25 cm depth) and compares these to the same characteristics of pellets of known species.

Nonmorphological tests were not very revealing. Hairs, which enter modern fecal pellets by being ingested when an animal licks itself or when it nibbles a bush another has rubbed against, were not found. We were not able to devise a chemical test to differentiate between the two types of pellets. Apparently fatty acids vaporize in the desiccation process. Endoparasites could not be found in the pellets (L.W. Dewhirst, University of Arizona, written communication, 1972). Enrichment broth, blood plates, and anaerobic cultures were run on the pellets by R.J. Trautman (University of Arizona, written communication, 1972) and the only organism found was *Bacillus* sp., an organism of soils. Fungal spores found in some of the pellets could have been ingested by the animals, or they could be post-depositional components.

In Table 11-3 of this paper are listed Long's radiocarbon dates of organic material from different levels in Stanton's Cave. The dates range from 1500 to ›35,000 years B.P. Presumably the base of the deposit is even older in the undated interval between 65 and 95 cm (Robert C. Euler, written communication, 1970); this interval contained only mashed pellets which could not be measured or otherwise analysed. The graph showing radiocarbon dates of pellets at different depths (Figure 11-6) indicates that slower deposition took place from 15,500 to 5760 years B.P. The age of the 20-25 cm interval, the interval in which fecal pellets of two different animals were found, averages 10,820 ± 140 years B.P. The large pellets, greater than 0.45 gm, and the small pellets, smaller than 0.22 gm, are not distinguishably different in age in this horizon. Therefore the different animals were at the cave during the same time interval.

The two methods used to mount reference plants were also used to mount plant matter from the pellets. Three to six pellets were examined from each 5 cm depth interval of Grid I-I, and, in all, 50 were examined in the intervals between 0 and 65 cm (Table 11-2). One half of each pellet was saved for further study or pollen analysis, and these were stored in the Geochronology Labs, University of Arizona.

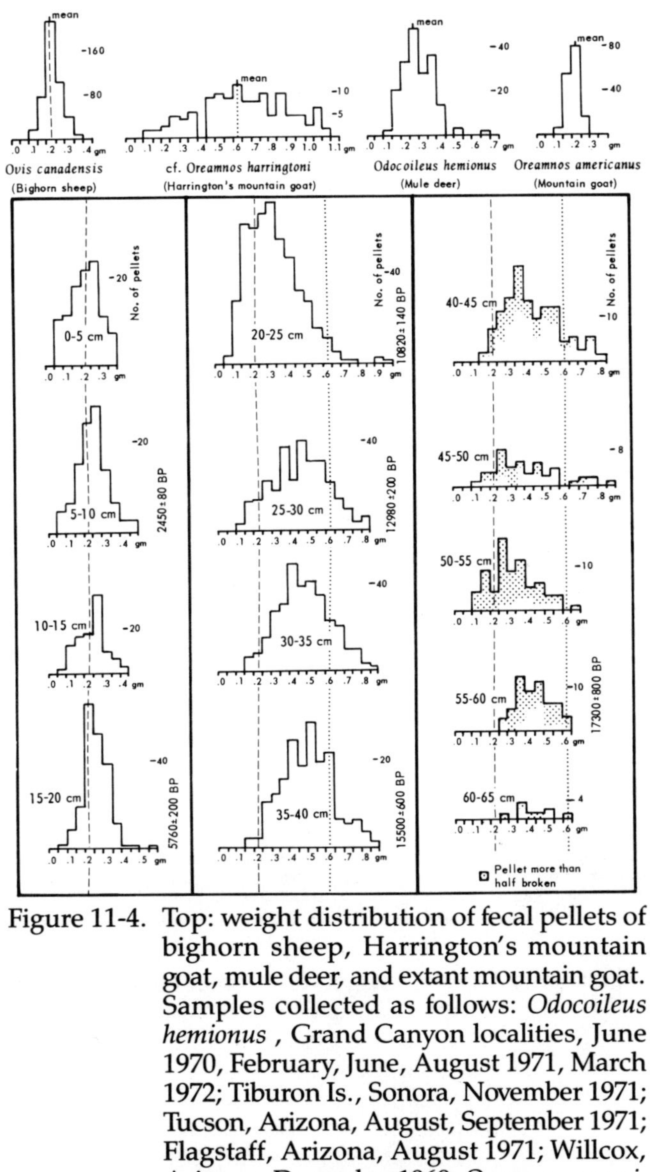

Figure 11-4. Top: weight distribution of fecal pellets of bighorn sheep, Harrington's mountain goat, mule deer, and extant mountain goat. Samples collected as follows: *Odocoileus hemionus* , Grand Canyon localities, June 1970, February, June, August 1971, March 1972; Tiburon Is., Sonora, November 1971; Tucson, Arizona, August, September 1971; Flagstaff, Arizona, August 1971; Willcox, Arizona, December 1969. *Oreamnos americanus,* Mt. Wardle, Kootenay National Park, Canada, June 1971. *Ovis canadensis,* Lake Mead, Arizona, March 1971; River Mts., Henderson, Nev., June to August 1970. cf. *Oreamnos harringtoni,* Rampart Cave, Arizona. Bottom: Artiodactyl fecal pellet weight distribution with age and depth, Stanton's Cave. Note significant reductions in weight distributions above 25 cm. Dashed line is mean weight of *Ovis canadensis.* Dotted line is mean weight of *Oreamnos harringtoni.*

We found that plant material in the fossil fecal pellets is preserved in a manner similar to that in pellets of living herbivores. Sheets of cells or cuticle, disassociated trichomes, and relatively low percentages of other items were identifiable in the fossil pellets. Plant fragments in pellets from the deeper intervals were even more decayed than those in pellets from the shallower intervals. In the lowest horizons, only plants having thick epidermal walls were identifiable. Pellets from the deepest levels had a whitish calcareous coating. Trampling, water (or urine), and time evidently promoted degradation of the pellets.

Robbins identified 89 species of vascular plants in the pellets. Photomicrographs of several of these fossil plant epidermal fragments are shown in Figure 11-7. Data sheets of plants identified in each pellet are available from the senior author. Table 11-1 provides both a listing of all plants identified in the pellets and an indication of where these plants presently are growing in the area. Botanical nomenclature follows that of Kearney and Peebles (1960). As shown in Table 11-2, fewer plant species were identified in pellets from the deeper, older intervals of Grid I-I than in pellets from shallower intervals. This decrease may be due to degeneration of plant tissue in the pellets or to a lower species diversity at the time the pellets were deposited (Hevly, this volume).

Several generalizations can be drawn concerning plant species identified in the pellets. 1) Of the 89 species, 28 are grasses (including rushes and sedges). 2) According to Kearney and Peebles (1960), 20 of the species found are indicators of a dry climate and six of continuously wet conditions. 3) Of the 50 pellets examined, 30% contained plants from a riparian habitat

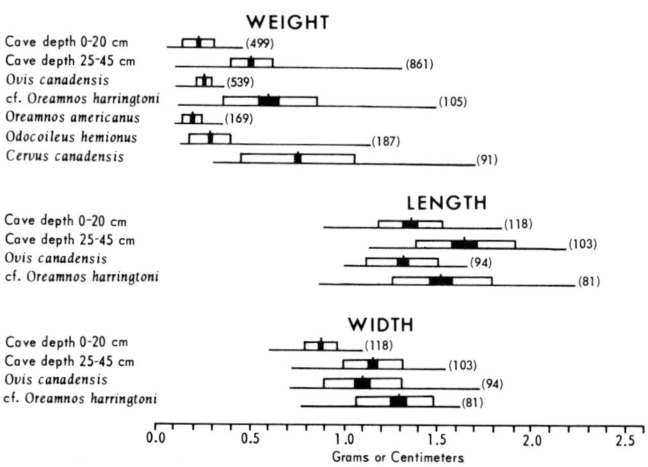

Figure 11-5. Graphic analysis and comparison of weight, length, and width of fecal pellets of bighorn sheep *(Ovis canadensis)*, Harrington's mountain goat (cf. *Oreamnos harringtoni)*, mountain goat *(O. americanus)*, mule deer *(Odocoileus hemionus)*, elk *(Cervus canadensis)*, and the pellets removed from Stanton's Cave. See Figure 12-4 for sample localities. *(Cervus* pellets collected by R.M. Hansen as follows: Cochetopa; Yellow Jacket Canyon, AZ., November 1974; Colo.; Jasper Nat. Park, Canada; Rocky Mt. Nat. Park; Guadalupe Mtns., Tex., Feb. 1974; Rock Springs, Wyo., 1974). Horizontal line indicates the range, vertical line the mean, open bar the standard error of the mean. (Method of Hubbs and Hubbs/1953). Number in parentheses indicates sample size.

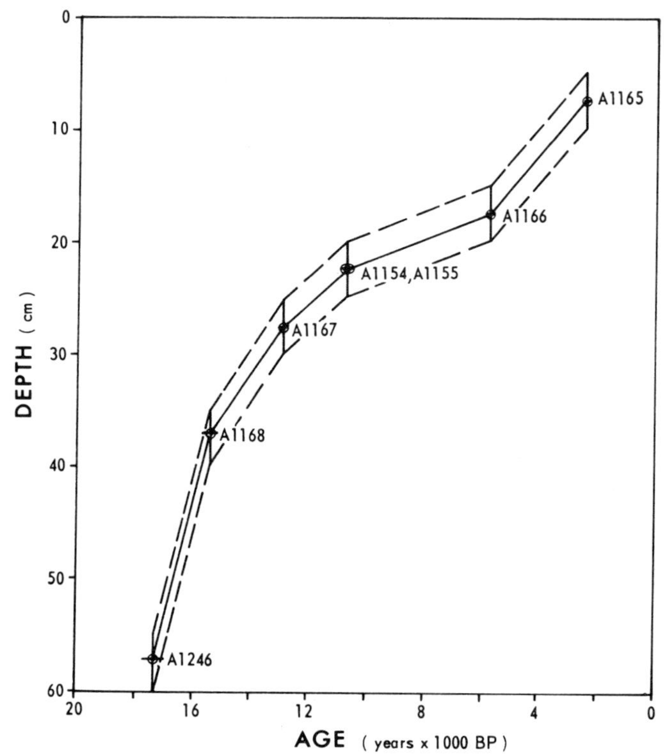

Figure 11-6. Arithmetic plot of radiocarbon dates versus depths in Stanton's Cave.

Figure 11-7. Photomicrographs of fossil epidermal fragments in pellets. Magnification: bar is 20 um.

1. *Aristida longisesta* 0-5 cm.
2. *Sporobolus* cf. *cryptandrus* 10-15 cm.
3. *Phragmites* sp. 30-35 cm.
4. Gramineae A 0-5 cm.
5. *Stipa arida* 0-5 cm.
6. *Adiantum* sp. 30-35 cm.
7. *Ephedra torreyana* 10-15 cm.
8. *Yucca angustissima* 50-55 cm.
9. *Artemisia tridentata* 0-5 cm.
10. *Coldenia hispidissima* 25-30 cm.
11. *Lesquerella* sp. 45-50 cm.
12. *Mentzelia* sp. 35-40 cm.
13. *Phoradendron* sp. 45-50 cm.
14. *Prunus fasiculata* 5-10 cm.
15. *Sphaeralcea* sp. 35-40 cm.
16. Fungus 50-55 cm.

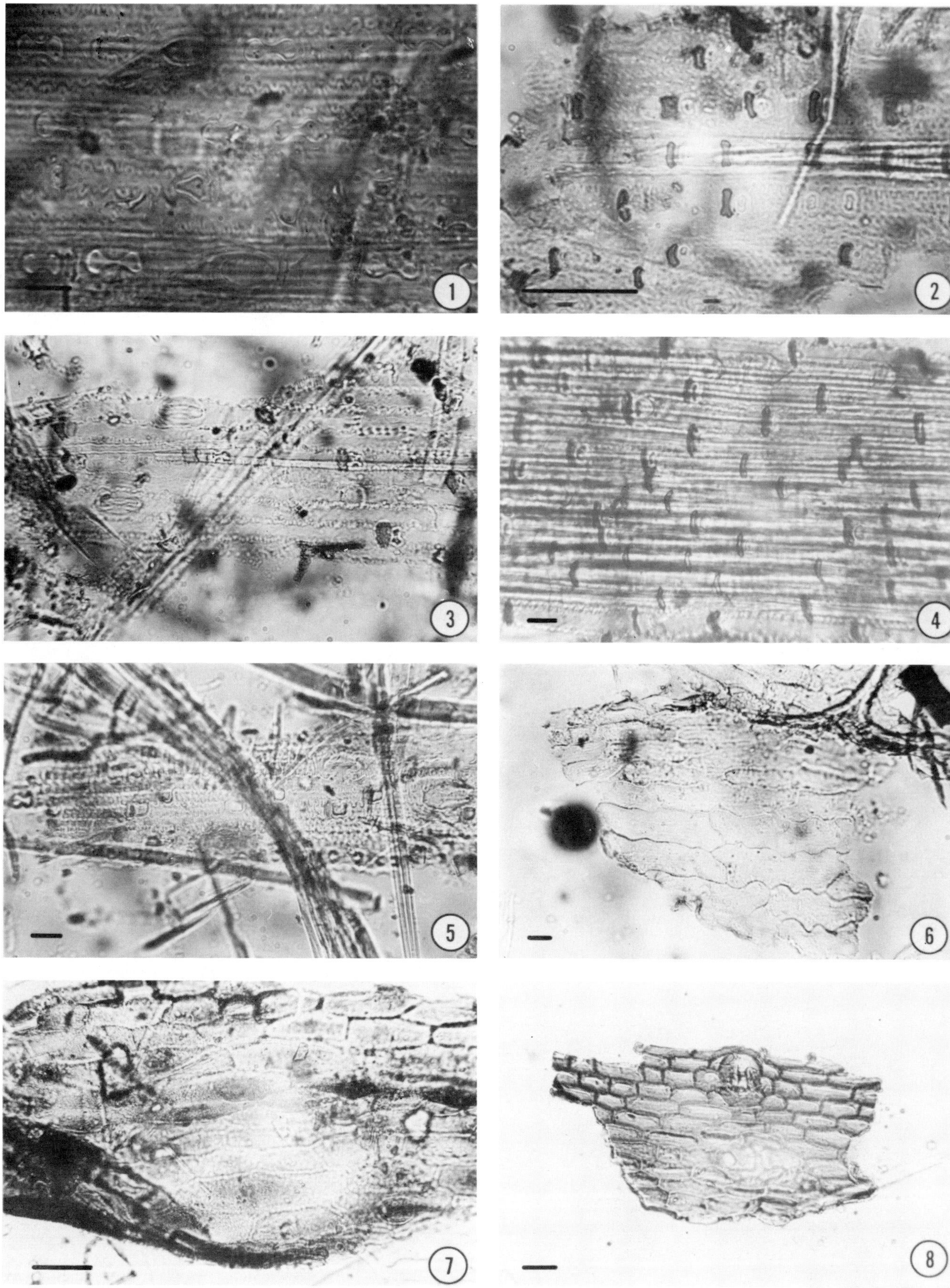
1
2
3
4
5
6
7
8

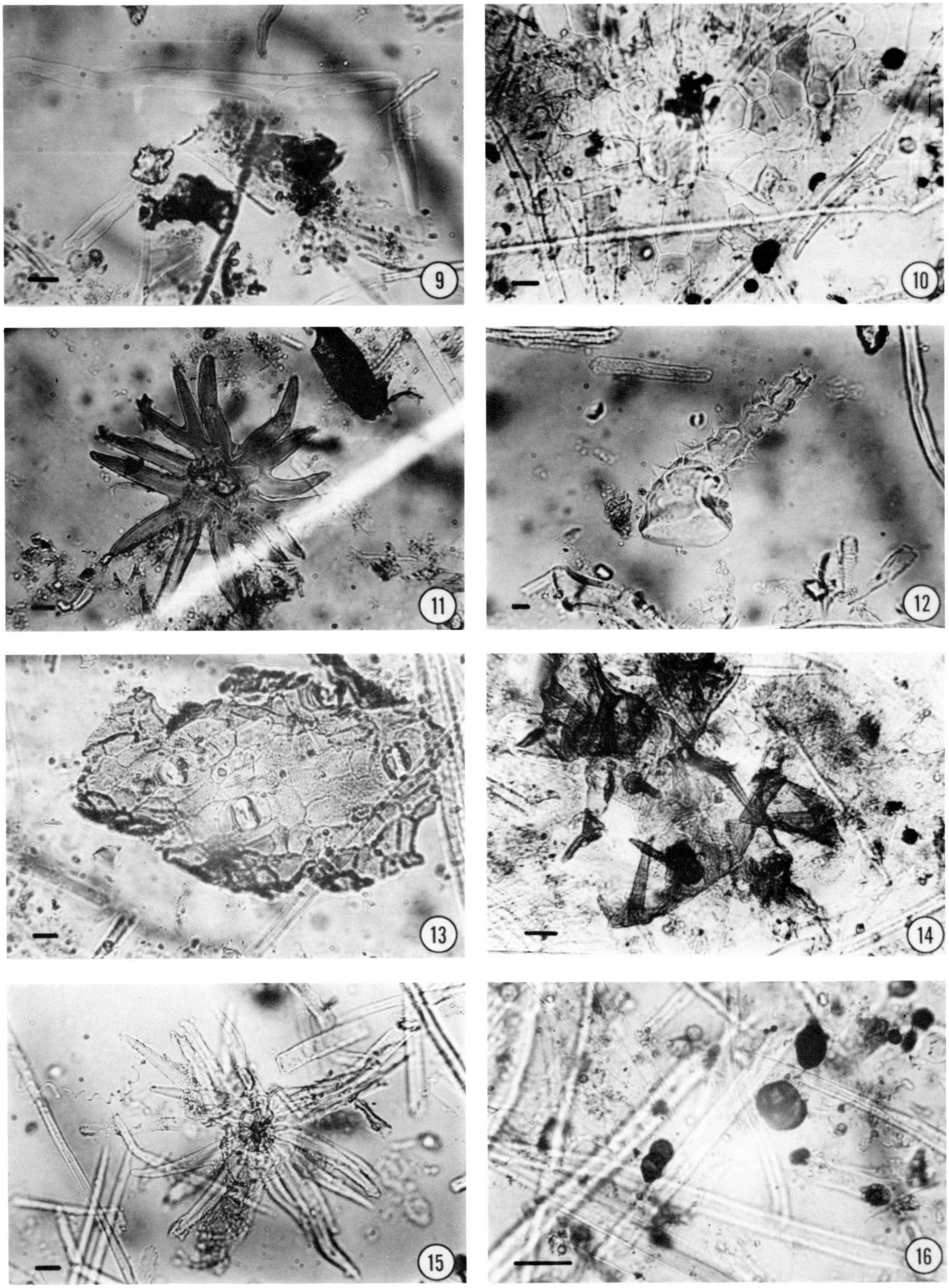
9
10
11
12
13
14
15
16

such as Vaseys Paradise or the area at river level, but no pellet contained only riparian taxa. 4) In the 0-25 cm interval, two indicators of dry conditions — *Bouteloua eriopoda* and *Ephedra* sp. — were found. In the same interval, *Carex* sp. is an indicator of a riparian habitat. In the 20-65 cm interval were found six indicators of dry conditions — *Aloysia wrightii, Bouteleoua* sp., *B. simplex, Coldenia hispidissima, Convolvulus* sp., and *Dyssodia pentachaeta*. Indicators for wet conditions found in this same interval are: *Clematis ligusticifolia, Eleocharis* sp., *Mentzelia* sp., and *Oenothera* sp.

We expected that a few plants would be eaten in such abundance that we could name them as dietary staples of the animals. Nine plants, six of them grasses, were identified in almost every interval, if not in every pellet. These are: *Agave* sp., *Agropyron* sp., *Aristida glauca, Enneapogon desvauxii, Panicum huachucae, Prunus fasiculata, Sporobolus* cf. *cryptandrus*, and an unidentified grass (Gramineae A in Table 11-1 and Figure 11-7.).

If vegetation levels were significantly depressed during the Pleistocene, we would expect to find in the pellets the same species which now grow at 1620-1890 m altitude in Kane Canyon; some of these plants that were identified in the pellets are *Arenaria* sp., *A. fendleri, Chrysopsis* cf. *hispida*, and *Yucca baccata*. *Arenaria* sp. could be *A. macradenia*, which today grows on dry slopes and ledges above 610 m (Kearney and Peebles 1960). Any other species of *Arenaria* would be an indicator of higher elevation. *Yucca baccata* grows along the Tonto platform and occasionally grows as low as the river level in Marble Canyon. *Chrysopsis* cf. *hispida* could quite possibly be growing now at the Stanton's Cave locality.

Analysis of Pollen in the Cave Earth

Martin analyzed 15 samples of cave earth for pollen. The samples were collected along the south wall of Grid I-I, and pollen was extracted following standard methods (Martin and Mehringer 1965). All samples proved rich in well-preserved pollen. Because the abundance of pine pollen masked other constituents, an initial 100-grain count determined the percentage of pine pollen in the samples. A second, 200-grain, count was then made ignoring pine pollen.

Martin recognized 45 pollen taxa. Depth profiles of the 15 most abundant pollen types are shown in Figure 11-8. Unknown pollen types constituted from a trace to 5% of each sample. Changes in pollen proportions are apparent in Figure 11-8. The proportion of *Artemisia* pollen increases markedly at 25 cm depth; this increase is associated with the appearance of small amounts of *Picea* and *Betula* pollen in the record. The proportion of pine pollen peaks at 25 cm below the surface of Grid I-I and varies more than that of any other pollen shown on Figure 11-8; the variation is probably caused by conditions on and above the rim.

Bartos (1972) extracted pollen from pellets collected in the 15-20 cm and 30-35 cm levels. In the pellets from the 15-20 cm level, she found less pine, juniper, and Cheno-Am pollen than in the cave fill. Bartos also found pollen from Rosaceae, Leguminosae, Nyctaginaceae, Ligulaeflorae, *Eriogonum*, and Euphorbiaceae, taxa whose pollen was not identified in the cave fill. In the 30-35 cm level, abundances of pollen in the pellets were similar to those in cave earth. However, Rosaceae was identified in pellets and not in the fill. In both pellet collections, pollen was not abundant, and Bartos (1972) concluded that the pellets were deposited in winter.

Fecal Pellets of Pleistocene Sheep and Goats

Modern species of mammals having fecal pellets similar in shape to the fossil ones include deer *(Odocoileus)*, elk *(Cervus)*, bighorn sheep *(Ovis)*, and mountain goat *(Oreamnos)*. Mule deer and bighorn sheep pellets are not easily distinguished (Murie 1954). Pellets of living mountain goats are significantly smaller than those of deer or sheep (Figure 11-3). Elk pellets are typically much larger than the fossil material from Stanton's Cave (Figure 11-5).

Although a bone from the cave has been questionably identified as a deer bone, we consider it unlikely that deer or elk could have descended the precipitous slopes into South Canyon to Stanton's Cave. Deer reach the Colorado River at Nankoweap (km 85.3) and move upstream, walking along terraces or swimming across the river, as far as Buck Farm Canyon (km 71) (Neil Guse, National Park Service, personal communication, 1972). Deer tracks have been seen at km 66 on the North

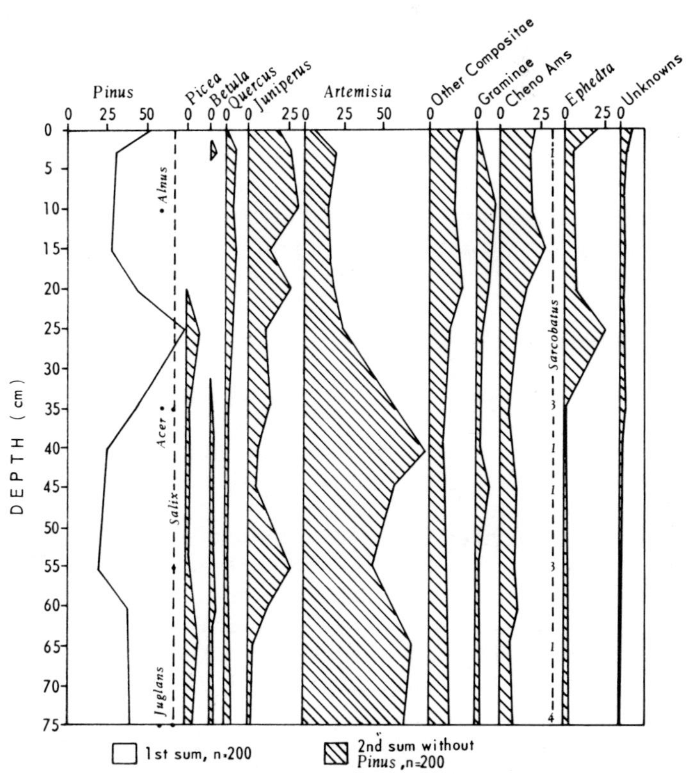

Figure 11-8. Pollen profile of cave earth, Stanton's Cave.

Rim and at km 63 on the South Rim. No elk bones have been identified from the cave.

Bighorn sheep have been sighted at the junction of the Colorado and Little Colorado rivers (km 65) since the time of Dellenbaugh (1887). As has been noted, two bighorn sheep were sighted near Redwall Cavern in August 1970 (Grand Canyon National Park records).

Figure 11-3 shows typical shapes and sizes of pellets from fossil localities and those from some modern species. Figure 11-5 compares length, width, and weight distributions of fossil and modern pellets. The range of weights of pellets from the upper 20 cm in Stanton's Cave (mean weight, 0.22 gm) overlaps the ranges of weights of pellets from modern mule deer (mean weight, 0.27 gm), bighorn sheep (mean weight, 0.22 gm), and mountain goat (mean weight, 0.19 gm). We conclude that the smaller pellets, dominant in the upper 20 cm of the deposit, were deposited by bighorn sheep *(Ovis canadensis)*. This conclusion is based on the pellet morphology, sightings of sheep in the vicinity, and the presence of bighorn sheep bones in the cave.

The larger pellets in the 25-65 cm interval are more difficult to identify. The mean weight is 0.45 gm. Similarly, the mean weight of similar shaped, handpicked pellets from Rampart Cave is 0.64 gm. Although the two are clearly not of the same population, the Rampart Cave collection mean is closer to 0.45 gm than that of any other pellet samples we have studied (see Figure 11-5).

Martin assigns the fecal pellets from Rampart Cave to Harrington's mountain goat, because its bones are second in number only to those of the Shasta ground sloth. These pellets are presently being studied at Colorado State University (R.M. Hansen, personal communication, 1977). We suspect that the large-pellet population from the lower levels in Stanton's Cave also represents Harrington's extinct mountain goat *(Oreamnos harringtoni)*. We know of no other herbivore, modern or fossil, that might be expected in Stanton's Cave and that might have fecal pellets having a mean weight of 0.45 gm.

Paleological Inferences

From the presence of two different types of artiodactyl pellets, we infer that at least two herbivores lived in the vicinity of Stanton's Cave during the past 18,000 years. The animals had to be able to climb steep slopes and shared cave-seeking behavior.

From the plant epidermal cells preserved in the smaller pellets, we conclude that the prehistoric bighorn ate a diet of 42% grasses and 58% herbs and shrubs. Bighorn living today in Arizona can consume a diet of as much as 75% grasses and sedges, the remainder being shrubs like *Ephedra*, *Physalis*, and *Prosopis* (Honess and Frost 1942; Jones 1950; Russo 1960).

The larger pellets, assigned to extinct Harrington's mountain goat, show the animals consumed a diet which averaged 25% grasses and 75% herbs and shrubs. Published reports for Cordilleran regions show extant mountain goats eat a diet of 63-96% grasses and sedges, the remainder being willow and mountain mahogany (Cowan 1940; Casebeer 1948; Brandborg 1955; Hibbs et al. 1969). Extant mountain goats have thick shaggy coats and normally inhabit cold, wet, mountainous, rocky areas where wind exposes Alpine grasses and shrubs (Brandborg 1955; Harington 1971). The extinct mountain goat occupied a different habitat, thriving for at least 7200 years on xerophytic vegetation.

Radiocarbon dates show that the animal having larger fecal pellets lived in the vicinity of the cave from 18,000 until 10,800 years ago. About 10,800 years ago, bighorn sheep appeared and persisted until 2000 years ago. Although occupation of the cave might not have been continuous, the radiocarbon dates show no obvious discontinuity.

That the cave was occupied seasonally is possible. Although no seasonality of visits to the cave is evident from plant remains in the pellets, other data suggest the pellets were deposited in winter or spring. Spring pellets in general have one flat end and one tailed end (Murie 1954). The Stanton's Cave pellets do not have these characteristics. On the basis of the total amount of pollen in a small representative collection of the pellets from Stanton's Cave, Bartos (1972) concluded that they were deposited in the winter. Forage is scarce along the rims of the Grand Canyon in the winter, and snowfalls can be intense. It is reasonable to propose that animals would migrate down into the inner gorge in winter to forage around Stanton's Cave and Vaseys Paradise and at times to take shelter in the cave.

The plant remains in the pellets provide information on both the diet of the animals and on the species of plants growing in the vicinity of Stanton's Cave during the past 18,000 years. The large pellets, from the 20-65 cm depth interval, indicate a diet of 75 species of plants, whereas the smaller pellets, from the 0-25 cm depth interval, contain 54 species. In the overlapping interval, 20-25 cm, 13% of the herbs and 25% of the grasses found in the large pellets were also found in the small pellets, indicating that both animals foraged on some of the same plants. No unexpected deficiencies exist in the list of plant species identified from the pellets. Even some annual species, which do not ordinarily survive digestion (Storr 1961), were identified. The occurrence of riparian species such as *Adiantum*, *Carex*, *Eleocharis*, *Equisetum*, *Mentzelia*, *Oenothera*, and *Phragmites* point to a former environment much like the present one, having occasional seeps or springs along the Canyon walls. All plant species identified in all the pellets presently grow within 19 km of the cave. Although bighorn sheep move an average of 5 km a day, under duress they could easily travel 19 km in one day (C.R. Hungerford, University of Arizona, personal communication, 1972). Therefore, the plant remains themselves may have come from various elevations and need not represent the vegetation immediately outside Stanton's Cave.

The changes in pollen percentages reveal climatic shifts. Pollen can enter cave earth from plants growing at various elevations in the area. In the Grand Canyon, the regional pollen is a complicated mixture (King and Sigleo 1973).

An increase in the proportions of pollen from *Artemisia, Betula, Picea, Pinus,* and other montane trees and shrubs in the 25-75 cm depth interval over those in the 0-25 cm depth interval indicates that the 25-75 cm interval appears to represent a time of cooler or wetter conditions and possible downward displacement of the woodland during the late Pleistocene. In the Stanton's Cave fill, a shift in dominance from pollen of *Artemisia,* which grows at high elevations where the conditions are cool and wet, to pollen of indicators of warm or dry conditions like Cheno-Ams and *Ephedra* took place around 13,000 years B.P. This climate change marks the end of the latest glaciation and evidence for it taking place at this time is also found in pollen records from Tule Springs, Nevada; Potato Lake, Arizona; San Augustin Plains, New Mexico; and Crane Lake, Texas (Martin and Mehringer 1965:440).

We are aware of the lack of evidence of vegetation change in the fecal pellets themselves. Possibly, grazing herbivores are less likely to ingest ecologically restricted plants. Or if the climate shift affecting the pollen preparations in the cave fill was a change in seasonality of rainfall, it might not have been intense enough to affect the plants eaten by the herbivores in this protected environment. Vaseys Paradise attracts riparian species limited by moisture rather than temperature. An important consideration is a 15° temperature difference between the warmer river level and the rim. Whatever pollen blows down from the rim reflects the vegetation changes occurring on the rim. Possibly, the plant community was adapted to annual spring flooding by the Colorado River, and now that flooding is no longer a yearly factor, the vegetation will shift to other elements previously stressed by flood.

Radiocarbon dating pinpoints the disappearance from Stanton's Cave of an extinct herbivore we assign to *Oreamnos harringtoni.* The 10,800 year-old date is within the range of other *O. harringtoni* dates in the Southwest (Harington 1971; see also Harington, this volume). *Nothrotheriops shastense* in the Grand Canyon (Long and Martin 1974) and *Mammuthus* sp. in other parts of the Southwest (Martin 1973) also became extinct about 10,800 years ago.

We conclude that: 1) The small pellets from Stanton's Cave represent bighorn sheep and the large pellets probably represent Harrington's mountain goat. Between 18,000-2,000 years B.P. they ate a variety of herbs and shrubs, all of which are growing in the area today. 2) While their diet as determined from plant epidermal fragments found in their fecal pellets did not change, pollen deposition in the cave did change appreciably. This pollen change represents either wetter or colder conditions on the rim or a change in seasonality of rainfall. 3) The disappearance of large pellets in the fill of Stanton's Cave 10,800 years ago represents the time of local extinction of Harrington's mountain goat. The time of local extinction coincides with the last record of Shasta ground sloths from Rampart Cave (Long and Martin 1974). This is the same or slightly later than the time of extinction of the mammoth *(Mammuthus* sp.), which Martin (1973) attributes to overkill by Paleoindians.

Acknowledgments

This paper is based on a Master's thesis written by the senior author at the University of Arizona (Iberall 1972). We express thanks to Robert C. Euler for support of the field work, to F.B. King for donating her time to the senior author as field assistant, to the National Park Service for authorizing permits, to Betty Fink for weighing pellets, and to the National Science Foundation for financial aid to P.S. Martin. Field excavations were supported by a National Geographic Society grant to Robert C. Euler and Prescott College.

Table 11-1. Plants identified in the fossil fecal pellets, Stanton's Cave.

(Nomenclature follows Kearney and Peebles, 1960. *Abbreviations:
SV — Stanton's Cave and Vaseys Paradise;
RP — rim above and along path to SV;
KC — Kane Canyon;
JL — Jacob Lake to Houserock;
OT — other localities.
The interrogation mark (?) is used to indicate that the identification is questionable.)

		Presence of plant cuticles in pellets (Depth from surface in cm)													
Taxa	Edaphic Significance	0-5	5-10	10-15	15-20	20-25	25-30	30-35	35-40	40-45	45-50	50-55	55-60	60-65	Occurrence*
Grasses															
Agropyron sp.		+	+	+	+	+	+	+	+	+	+	+	+		SV
Andropogon sp.						+			+						OT
Aristida sp.	dry				+	+	+		+						SV
A. longiseta		+	+		+	+	+	+		+	+				JL
A. glauca	dry	+	+	+	+	+		+	+	+	+		+	+	SV
Bouteloua sp.	dry						+	+							SV
B. eriopoda	dry	+	+												OT
B. gracilis	dry			+							+				SV
B. simplex	dry						+		+						OT
B. trifida	dry	+	+		+		+	+							OT
Bromus sp.						+		+							OT
Enneapogon desvauxii		+	+	+	+	+	+	+	+		+	+			JL
Festuca sp.		+		+	+					+					OT
Lycurus phleoides		+													OT
Oryzopsis sp.			?		+			?	?						OT
Panicum huachucae		+	+	+	+	+	+	+	+	+	+		+	+	SV
Schedenardus paniculatus		+	+	+				+	+	+	+				OT
Sporobolus sp.	some dry	+	?					+				+			JL
S. cf. cryptandrus	dry	+	+	+	+	+	+	+	+	+	+		+		JL
S. flexuosus				+	+		+								OT
S. texanus		+	+	+	+	+	+	+	+	+					OT
Stipa arida	dry	+	+				+	+	+	+	+				SV
Tridens pulchellus		+	+		+		+	+	+	+			+		RP
Gramineae A		+	+	+	+		+	+	+	+	+	+	+		OT
Unknown grasses			+				+	+	+	+	+			+	OT
Rushes and sedges															
Carex sp.	wet	+	+												SV
Eleocharis sp.	wet										+				SV
Equisetum sp.	moist	+					+								SV
Phragmites sp.	moist	+	+				+	+	+	+	+		+		SV

Taxa	Edaphic Significance	Presence of plant cuticles in pellets (Depth from surface in cm) 0-5	5-10	10-15	15-20	20-25	25-30	30-35	35-40	40-45	45-50	50-55	55-60	60-65	Occurrence*
Herbs and trees															
Adiantum sp.	moist				+			+							SV
Agave sp.		+	+	+		+	+	+	+	+		+	+		SV
Agave utahensis							+								SV
Aloysia (Lippia) wrightii	dry							+	+						SV
Amelanchier sp.		+													OT
Aplopappus sp.				+											OT
Arenaria sp.		+			+		+					?			KC
Arenaria fendleri									+						KC
Artemisia tridentata		+	+			+	+	+	+	+	+		+		RP
Aster sp.	dry	+	+		+			?		+					SV
Astralagus sp.		?	+	+		+	?	+	+		?	?			SV
Atriplex sp.		+	+	+	+	+	+	+	+	+	+	+	+		RP
A. cf. jonesii		+													OT
Baccharis sergiloides							+								SV
Berberis repens					+										OT
Brickellia sp.			+												SV
Cercocarpus intricatus								+							RP
Chenopodium sp.				+					?						OT
Chrysopsis cf. hispida						+									KC
Chrysothamnus sp.	dry	+	?											+	SV
Clematis ligusticifolia	streams								+	+	?				RP
Coldenia sp.	dry	+	+	?	+			+	+	+	+				SV
C. hispidissima	dry						+								RP
Coldenia-Cryptantha	dry		+				+		+	+	+	+			SV
Convolvulus sp.	dry								+						KC
Cryptantha sp.	desert	+	+			+			+	+	+	+			SV
Datura metaloides		+													OT
Dyssodia sp.										+					SV
D. pentachaeta	dry									+					SV
Ephedra sp.	dry					+									SV
E. torreyana	dry		+	+								+	+		SV
Eriogonum sp.	dry/washes	?	+				+	+	?	?			?	?	SV
E. deflexum	dry		+		+				+					?	RP
Euphorbia sp.							+								SV
Fallugia paradoxa			+					+							SV
Galium sp.		+	+			+	+	+	+	+					RP
Gutierrezia sp.	dry			+			+								SV
Hedeoma sp.							+								SV
H. diffusum								+							SV
Laphamia congesta	dry		?	+						+	+	+	+		SV
Lesquerella sp.				?							+				KC
Lygodesmia exigua										?	+				SV
Mentezelia sp.	washes					+			+			+			SV
Mimulus sp.	streams	+				+	+	+		+	+			+	SV

Taxa	Edaphic Significance	Presence of plant cuticles in pellets (Depth from surface in cm) 0-5	5-10	10-15	15-20	20-25	25-30	30-35	35-40	40-45	45-50	50-55	55-60	60-65	Occurrence*
Oenothera sp.	washes						+								SV
Opuntia basilaris									+					+	SV
Phoradendron sp.							+				+				OT
Physalis fendleri										+		?			JL
Prunus fasiculata	dry	+	+	+	+	+	+	+	+	+	+				SV
Rhus trilobata										+					SV
Sarcobatus vermiculatus							+								OT
Solidago sp.	some dry		?							+					SV
Sphaeralcea sp.		+		+	+	+	+	+	+	+	+				SV
Tetradyma canescens?							+	+		+		+		+	SV
Tragia sp.										+					SV
Urtica serra										+					SV
Yucca sp.		?						+				+	+	+	SV
Y. angustissima						+		+	+		+	+	+	+	SV
Y. baccata		+			+				+						KC
Fungal spores		+			+							+			

Table 11-2. Summary of data from fecal pellets removed from Stanton's Cave.

Depth in cm	Number of pellets removed from meter trench	Number of pellets analyzed for epidermal fragments	Number of species of plants identified in fecal plants
0-5	153	5	41
5-10	242	3	37
10-15	153	3	24
15-20	711	5	23
20-25	1114	3	23
25-30E	897		
25-30W	441	6	36
30-35	450	3	35
35-40	205	5	34
40-45	121 (33 whole)	3	35
45-50	44 (19 whole)	3	25
50-55	71 (7 whole)	3	18
55-60	53 (5 whole)	3	17
60-65	9 (2 whole)	5	9
Total	4664	50	

Table 11-3. Radiocarbon dates from Stanton's Cave.

A—1,000 series samples dated at University of Arizona Isotope Dating Laboratory. Dates from pellets unless otherwise noted.

Isotope Dating Lab Number	Date Measured (B.P.)	Location and Comments
A1184	1,500	*Pinus edulis* driftwood on surface in back of cave.
A1165*	2,450 ± 80	5-10 cm; Grid I-I.
Euler & Olson, 1965	4,095 ± 100	Split-twig figurines.
A1166*	5,760 ± 200	15-20 cm; Grid I-I.
A1154*	10,760 ± 200	Small fecal pellets (‹.25 g); 20-25 cm; Grid I-I.
A1155*	10,870 ± 200	Large fecal pellets (›.5 g); 20-25 cm; Grid I-I.
A1167*	12,980 ± 200	25-30 cm; Grid I-I.
A1082	13,070 ± 470	20-25 cm; Grid AA.
A1132	13,770 ± 500	Large fecal pellets; 20-25 cm; Grid GG.
A1238	15,230 ± 240	*Teratornis* bone. Not *in situ*.
A1168*	15,500 ± 600	35-40 cm; Grid I-I.
A1246*	17,300 ± 800	55-60 cm; Grid I-I.
A1056	›35,000	Driftwood at base of section below 65 cm; Grid I-I, east trench. (See also Elston, Chapter 11, this volume.)

*Data used in Figure 11-6.

References

Bartos, F.M.
1972 Fecal pellets as environmental indicators. Unpub. Master's thesis. University of Arizona. 96 pages.

Brandborg, S.M.
1955 *Mountain Goat in Idaho.* Idaho Dept. Fish Game Wildlife Bull. 2. 142 pages.

Casebeer, R.L.
1948 A study of the food habits of the mountain goat *(Oreamnos americanus missoulae)* in western Montana. Unpub. Master's thesis. Montana State University. 85 pages.

Cowan, I.M.
1940 Distribution and variation in the native sheep of North America. *Am. Midland Naturalist* 24:3:505-580.

Croker, B.H.
1959 A method of estimating the botanical composition of the diet of sheep. *New Zealand Jour. Agri. Research* 2:72-85.

Dellenbaugh, F.S.
1887 The great walled river. *Am. Geog. Soc. Bull.* 19:2:154 pages.

Euler, Robert C., and A.P. Olson
1965 Split-twig figurines from northern Arizona: new radiocarbon dates. *Science* 148:3668:368-369.

Hansen, R.M.
1971 *Drawings of tissues of plants found in herbivore diets and in the litter of grasslands.* Grassland Biome, U.S. Internat. Biol. Program Tech. Rept. 70, 69 pages.

Harington, C.R.
1971 A Pleistocene mountain goat from British Columbia and comments on the dispersal history of *Oreamnos. Canadian Jour. Earth Sci.* 8:9:1081-1093.

Hibbs, D., F.A. Glover, and D.L. Gilbert
1969 The mountain goat in Colorado. *North Am. Wildlife Nat. Resource Conf. Trans.*, pp. 409-418.

Hoffmeister, D.F.
1971 *Mammals of Grand Canyon.* Chicago: University of Illinois Press, 183 pages.

Honess, R.F., and N.M. Frost
1942 *A Wyoming bighorn sheep study.* Wyo. Game and Fish Dept. Bull 1, 127 pages.

Hubbs, C.L, and C. Hubbs
1953 An improved graphical analysis and comparison of series of samples. *Systematic Zoology* 2:49-56.

Iberall, E.R.
1972 Paleoecological studies from fecal pellets: Stanton's Cave, Grand Canyon, Arizona. Unpub. Master's thesis, University of Arizona, 67 pages.

Jones, F.L.
1950 *A survey of the Sierra Nevada bighorn.* Sierra Club Bull, June 1950, 76 pages.

Kearney, T.H., and R.H. Peebles
1960 *Arizona Flora,* 2nd ed. Berkeley: University of California Press, 1032 pages.

King, J.E., and W.R. Sigleo
1973 Modern pollen in the Grand Canyon. *Geosciences and Man* 7:73-81.

Laudermilk, J.D., and P.A. Munz
1938 Plants in the dung of *Nothrotherium* from Rampart and Muav caves, Arizona. *Carnegie Inst. Washington Pub.* 487:271-281.

Long, Austin, R.M. Hansen, and P.S. Martin
1974 Extinction of the Shasta ground sloth. *Geol. Soc. America Bull.* 85:1843-1848.

Long, Austin, and P.S. Martin
1974 Death of American ground sloths. *Science* 186:4164:638-640.

Madsen, D.B.
1973 Late Quaternary paleoecology in the Southeastern Great Basin. Unpub. Ph.D. dissertation, Univ. Missouri, 125 pages.

Martin, D.J.
1954 Features of plant cuticle: an aid to the analysis of the natural diet of grazing animals, with especial reference to Scottish hill sheep. *Trans Proc. Bot. Soc. Edinb.,* 36:278-288.

Martin, P.S.
1973 The discovery of America. *Science* 179:4077: 969-974.

Martin, P.S., and P.J. Mehringer
1965 Pleistocene pollen analysis and biogeography of the Southwest." *In* Wright, H.E., Jr., and D.G. Frey, eds., *The Quaternary of the United States.* Princeton, N.J.: Princeton Univ. Press. pp. 433-451.

Martin, P.S., B.E. Sabels, and Dick Shutler, Jr.
1961 Rampart Cave coprolite and ecology of the Shasta ground sloth. *Am. Jour. Sci.* 259:2:102-127.

Murie, O.J.
1954 *A field guide to animal tracks.* Boston: Houghton Mifflin Co., 374 pages.

Russo, J.D.
1960 *The desert bighorn sheep in Arizona.* Ariz. Game and Fish Dept. Wildlife Bull. 1, 153 pages.

Sparks, D.R. and J.C. Malecheck
1968 Estimating percentage dry weight in diets using a microscopic technique. *Jour. Range Management* 21:264-265.

Stace, C.A.
1965 Cuticular studies as an aid to plant taxonomy. *British Mus. (Nat. History) Bull., Botany* 4:1:78 pages.

Stewart, D.R.M., and Joyce Stewart
1970 Food preference data by fecal analysis for African plains ungulates *Zoo. Africana* 5:1:115-129.

Storr, G.M.
1961 Microscopic analysis of faeces, technique for ascertaining the diet of herbivorous mammals. *Australian Jour. Biol. Sci.* 14:157-164.

Van Devender, T.R., and J.E. King
1971 Late Pleistocene vegetational records in Western Arizona. *Ariz. Acad. Sci. Jour.* 6:4:240-244.

Welles, R.E., and F.B. Welles
1961 *The bighorn of Death Valley.* National Park Service Fauna Ser. 6, 242 pages.

Wells, P.V.
1966 Late Pleistocene vegetation and degree of pluvial climatic change in the Chihuahuan Desert. *Science* 153:3739:970-975.

Wells, P.V., and Rainer Burger
1967 Late Pleistocene history of coniferous woodland in the Mohave Desert. *Science* 155:3770: 1640-1647.

Wells, P.V., and C.D. Jorgensen
1964 Pleistocene wood rat middens and climatic change in Mohave Desert: a record of juniper woodlands. *Science* 143:3611:1171-1174.

Wilson, R.W.
1942 Preliminary study of the fauna of Rampart Cave, Arizona. *Carnegie Ins. Washington Pub. 530* 6:169-185.

Chapter 12

Stanton's Cave During and After the Last Ice Age

by

Paul S. Martin
Department of Geosciences
The University of Arizona, Tucson

In northern Arizona below Lees Ferry the Colorado River enters the Grand Canyon. Within the next 70 miles (113 km) the Colorado exposes successively older rocks, which rise as cliffs or precipitous slopes to embrace the river ever more deeply. Thanks to the flexure of the Echo Cliffs Monocline, indolent river rafters are able to examine at leisure the unfolding of the geological formations. In order of increasing age these are massive walls of Kaibab and Toroweap limestones; the Coconino Sandstone; the Hermit Shale (soft, brick red and sparsely vegetated); the steep-sloping Supai Formation; and by river mile 22.6 (km 36.4) below Lees Ferry the top of the cliff-forming, cave-rich Redwall Limestone. Unfortunately for prospectors of ice age fossils the 30 river miles (48 km) in which the Redwall is readily accessible at or near river level are not notably rich in dry caves. Stanton's Cave (Figure 12-1) is an exception. Through the lower 240 miles (386 km) of the Grand Canyon the Redwall lies above, usually far above, the Colorado River. The gaping mouths of its many caves are best known to bats and owls.

In 1869, a little over 30 river miles (48 km) from Lees Ferry, John Wesley Powell came upon "fountains bursting from the rock," a cascade from small cave holes in the Redwall. The seepage supports a verdant patch of monkey flower, columbine, Emory broom, poison ivy and an occasional orchid. Powell named the garden Vaseys Paradise for his botanist friend George Vasey; a more desirable memorial is hard to imagine. If Powell's party explored the large, conspicuous dry cave mouth in the Redwall limestone a few hundred meters upstream from Vaseys Paradise, he neglected to mention it.

There, within the soft rose-tinted light of a large room littered with limestone blocks detached from the ceiling, Robert Stanton stored his gear after injury and drowning ended his disastrous Colorado River expedition of 1889. Many years later, the cave which sheltered Stanton's equipment was discovered to contain many unusual prehistoric artifacts known as split-twig figurines, leading to Robert C. Euler's expedition sponsored by the National Geographic Society.

Figure 12-1. Aerial view of Stanton's Cave.

The natural vegetation at this end of the Canyon is derived mainly from the Great Basin, gradually giving way downstream to Mohave Desert species. Catclaw *(Acacia greggii)* and mesquite *(Prosopis torreyana)* reach their upstream limits at mile 39 (km 63) and mile 40 (km 64), respectively. The sequence of other important Mohave Desert species as they first appear below Lees Ferry include: brittle bush *(Encelia farinosa)* 45 miles (72 km); Mormon tea *(Ephedra nevadensis)* 61 miles (98 km); white bursage *(Fransera dumosa)* 134 miles (216 km); ocotillo *(Fouquieria splendens)* 154 miles (248 km); and creosote bush *(Larrea tridentata)* 170 miles (274 km). One thousand feet (305 m) lower in elevation, the desert community adjacent to the river in the western end of the Canyon is much richer in woody plant species than the gorge of Marble Canyon around Stanton's Cave. Woody plants found on the terrace slopes near Stanton's Cave include Indian tea *(Ephedra torreyana)*, golden weed *(Haplopappus)* and mescal *(Agave utahensis)* with broom-like *Brickellia longifolia*, tamarisk, *Baccharis sergiloides* and occasional seedlings of redbud *(Cercis)* on riparian habitat near the river. Before construction of Glen Canyon Dam the riparian border of the Colorado River regularly reached a high flood level still marked in this end of the Canyon by occasional large bushes of Apache plume *(Fallugia)*. Woody plant coverage below Stanton's Cave, as elsewhere in most of the Inner Gorge, is sparse — about 15% or less.

Of course Powell did not discover Vaseys Paradise anymore than Stanton discovered the cave that came to be named for him. Both were known to prehistoric people thousands of years earlier, people who made their way down from House Rock Valley into the Inner Gorge, to leave their offerings in the great room above the river. That offering was the simple yet exquisite toy-like image of a deer or sheep whose radiocarbon age is consistently found to be around 4000 years. A figurine can be constructed in half an hour by a dexterous craftsman, a copy that lacks only the weathered patina of the originals. Just as any attractive ancient stone artifact tempts its discoverer to remove it for some undetermined purpose, so after 4000 years of sanctuary the twig figurines were being removed from Stanton's Cave. Alert to the importance of context, to the need for salvage, and especially to the paleoecological opportunity, Bob Euler led two helicopter expeditions to Stanton's Cave in search of ancient human remains, and any other evidence of past life that would help illuminate the ancient environment of America's most famous canyon. Except for finding and mapping many more figurines lying on the cave surface or shallowly buried, the archeological results were sparse, and thus similar to other Grand Canyon cave finds (e.g., Schwartz et al. 1958). The cave earth itself was shallow, a meter or less of fill overlying roof fall. However, for those interested in ancient environments certain fossils uncovered by Euler in the fill beneath the figurines were very promising. Among other paleoecological treasures Euler had found the perishable remains of an extinct goat.

Spectacular, a word that fits the scenery outside Stanton's Cave, also applies to the fossils within. Unlike gold in Tutankhamen's Tomb, their value may not be instantly apparent. However, for anyone who would investigate the diet of the large mammals that roamed North America until remarkably late in the ice ages, and for those who would seek to accomplish the seemingly impossible feat of learning the life history of extinct mammals, the fossil dung, horn sheaths, and other perishable materials of the extinct beasts are treasure indeed. From such carcass remains one obtains direct radiocarbon dates on when the animal in question lived, and, by judicious inference, when it did not. For radiocarbon dating of an extinct mammal there is no better material than horn sheaths, hair or dung.

From plant cuticles in well-preserved ancient feces, one recovers a fossil plant record of what the extinct animal ate. Fossil pollen and seeds in cave earth adjacent to bone and dung deposits yields information of any major changes in climate. From the assembled paleoecological dates one may gain a picture of the natural environment of the Grand Canyon through the last ice age, the diet of the extinct animal, and the other plants and animals that were its ancient associates (see Robbins et al., this volume).

Typically geologists visit the Grand Canyon in search of much older fossils and much older events. In the chronology of the Grand Canyon a few millenia may slip by like a boat over Lava Falls, come and be gone before the full significance of the event can be appreciated. Few visitors to the Grand Canyon expect to learn about plant migrations at the end of the last ice age or to encounter evidence of extinct animals of "only" 11,000 years ago. Even fewer might expect to learn in detail about the diet of the missing mammals. To gain such information one needs dung and other perishable remains. How can the spoor of an extinct animal be identified?

Two-hundred and forty miles (386 km) downstream from Stanton's Cave at the other end of the Canyon is Rampart Cave, another remarkable tomb. It contained a rich stratified deposit of dry dung, hair, and other perishable parts of the Shasta ground sloth *(Nothrotheriops shastensis)*, as well as bones of the ground sloth, Harrington's extinct mountain goat *(Oreamnos harringtoni)* and horse *(Equus)*. The deposit of softball-size lumps of dung ranks in paleoecological significance with the frozen mammoth cemeteries of Siberia. Tragically, a fire recently destroyed two-thirds of the contents of the cave. The second most common large mammal in Rampart Cave, after the Shasta ground sloth, was Harrington's extinct mountain goat. According to Wilson (1942), "... skeletal elements are so common in the small Test Pit B that many of these animals must have inhabited the area at the time the cave was occupied." While Wilson also reported horn sheaths of *Oreamnos*, he did not mention unusual fecal pellets, which were collected during the extensive and largely unpublished excavations of Rampart Cave by Remmington Kellog. The pellets, roughly the size of a cough drop and up to 0.7 grams in dry weight, were far too small and fine textured to represent the dung of even a small ground sloth. They were larger than fecal pellets of living mountains goats, mountain sheep, or deer (Murie 1974).

When Bob Euler invited me to visit Stanton's Cave in 1969, he reported that pellets were more common than any other item in his excavations. The short climb to the cave led to an unforgettable encounter with fine cave earth being sampled and screened by Euler's crew. On the screens were fecal pellets in abundance. They came in both large and small sizes. No one, to my knowledge, had reported such variation before. The pellets occurred throughout the upper half meter of cave fill, and were concentrated at a depth of 25 cm, as disclosed by Euler's stratigraphic sampling of various test pits. Pellets were by far the most common item in the cave earth. Their meaning began to suggest itself.

Long before ancient man arrived, the cave was a refuge to native herbivores of the ice age, animals that perhaps appreciated the lush growth around Vaseys Paradise even more than John Wesley Powell. As the report by Robbins et al. (this volume) shows, about half of the fossil pellets averaged one quarter of a gram in weight. These were small enough to be within the size range of droppings of mountain sheep or mule deer, animals to be expected at or near the cave today. The rest were too large. They weighed half a gram or more and, while not identical in shape, they were similar in weight to those found at Rampart Cave and attributed to Harrington's mountain goat (Figure 12-2). Given evidence of large and small pellets at Stanton's Cave, it was no surprise that the fossil bones included those

Figure 12-2. Profile of wall of Grid I-I showing artiodactyl dung pellets. North arrow is 30 cm long.

of both mountain sheep and extinct goat (see Harrington, this volume). The collection also yielded twelve horn sheaths of Harrington's extinct goat. Barring an occasional mountain lion, sheep and extinct goats were the largest mammals to live inside the cave before people came.

The identification of the large pellets is based on the observation that they are only known from caves also harboring bones or distinctive horn sheaths of the extinct goat, *Oreamnos harringtoni*. Besides Rampart and Stanton's caves, extinct goat bones and large pellets were found at Bida Cave (Mead 1983). In each case the large pellets are larger than deer, mountain sheep or living mountain goat droppings and subequal in size to elk. Since no bones of large cervids have been found in Grand Canyon caves, I think there is little chance the dung was deposited by elk. Admittedly the pellets from Stanton's Cave are not identical in shape with those from Rampart Cave. Furthermore, its bones indicate Harrington's goat was smaller, not larger than living mountain goats which void pellets even smaller than mountain sheep (Robbins et al., this volume). However, the repeated association of distinctively large fecal pellets with bones of Harrington's extinct goat is highly suggestive.

From recent accelerator dates on the carbon fourteen (^{14}C) content of their horn sheaths, we know that Harrington's goat lasted until 11,000 years ago (see Table 6-1). Dates on the large fecal pellets show that they were being voided in Stanton's Cave until 11,000 years B.P. or perhaps slightly later. The disappearance of both goat horns and mystery pellets agrees with radiocarbon dates on extinction of Shasta ground sloths in the western end of the Canyon and in the Guadalupe Mountains of west Texas (Thompson et al. 1979). Provocatively, the dates are also coeval with the time when mammoth bones in southern Arizona are found associated with the Clovis points and other artifacts of American's first big game hunters, suggesting the possibility of a common cause in the extinction of goats, ground sloths and mammoths.

The adjective spectacular again seems appropriate in describing the fossil birds of Stanton's Cave, such as the first fossil record of a tropical robin native to northeastern Mexico, *Turdis grayii*. The fossils also include a cranium and humerus of *Teratornis merriami*, an extinct vulture larger than any living vulture including the California condor. In the Serengeti area of East Africa, there are seven species of vultures which may gather in distinct combinations to feast on a carcass (Houston 1975). A similar fauna of large bird including the teratorn, the condor, an extinct black vulture and other extinct vultures apparently gathered to dine on the hearty meals offered up in the Pleistocene at the La Brea tar seeps. In the Grand Canyon the radiocarbon date on the teratorn humerus and pollen from the bones of *Teratornis* show that it lived during the time of the extinct goats.

Might the condors or teratorns have survived in the Grand Canyon thousands of years after Harrington's mountain goats, the Shasta ground sloths, burro-sized equids, and other extinct fauna vanished? Condor feathers of undetermined age have been found on the surface of certain dry caves. However, in an environment where extinct goat feces survived at least 19,000 years and sloth dung for over 40,000 years, good preservation alone is clearly not sufficient proof of a sub-Recent age. If some inaccessible cave in the Redwall yields to the ropes of rockclimbers and more condor or teratorn bones, or even feathers and eggs are found together in an ancient nest, I predict that any radiocarbon dates on the preserved organic fraction, no matter how fresh it may appear, will be older than 10,000 years.

My expectation is based on other radiocarbon samples of very youthful materials, materials of the highest quality for radiocarbon dating, such as those from Stanton's Cave. No such dates have been presented to seriously challenge the 9th millenium B.C. as the fatal time for the last of the Pleistocene megafauna, teratorns included. Every ^{14}C date on the remains of extinct fauna offers a fresh test of this view and, given the volatile nature of negative evidence, such test opportunities should be seized.

In my enthusiasm for recounting the discovery of extinct animal remains, I have slighted other paleoecological treasures from Stanton's Cave. "Spectacular" may not be too strong an adjective to apply to a deposit yielding quantities of ancient driftwood, including logs of Douglas-fir and poplar whose rings can be cross-dated, if not calendar dated. The wood recovered by Euler from his first Stanton's Cave expedition proved to be >42,000 years B.P. in radiocarbon years (Figure 12-3). Wes Ferguson has discovered that the piñon in the modern beach rack may be thousands of years old. What might account for the much older driftwood wedged in boulders below the goat pellets in Stanton's Cave?

In historic time before power dams intervened, a good spring flood on the Colorado surged to ten times the flow that Powell encountered. Since Glen Canyon Dam, river fluctuations have been converted from an annual to a daily event, regulated by the voracious appetite for power of the southwestern cities which peaks in late afternoon. The greatest of the recorded pre-dam discharges, approximately 300,000 cfs at Lees Ferry, may have put the driftwood in the cave (See Hereford, this volume).

A higher Pleistocene discharge of the Colorado is suggested by bouldery terraces from side canyons downstream, which were deposited 80 feet (24 m) or more above the historic maximum floodline. Perhaps a deluge on a side tributary managed to deposit a massive boulder fill that dammed the Colorado at a time when the main stem itself lacked the energy to clear its channel. The mystery of fossil driftwood is enhanced by the fact that the radiocarbon age predates the last glacial maximum of 14 to 24 thousand years ago, a time when I would expect the main opportunity for extreme run-off and maximum river discharge. While the hydrological meaning of the driftwood remains a mystery, the ecological implications are clear. Piñon pine, one of the dominant trees contributing to the modern driftwood rack along the river, was not found among the ancient fossil wood in Stanton's Cave. The upstream environment must have been unfavorable for piñon when the driftwood floated into the cave. Upstream, other montane conifers had replaced today's abundant piñon-juniper woodlands, as they did in the eastern Grand Canyon in Cole's (1982) record of fossil pack rat middens.

The opportunity for more driftwood discoveries along the river should not be overlooked. In such a dry environment, old driftwood may last on terraces for thousands of years, with or without the benefit of protection by a cave or rock shelter. When protected it obviously will last much longer. The ancient driftwood in Stanton's Cave led to another thought. Any prehistoric or, for that matter, historic people entering the cave might well have converted the ancient wood to fuel. An unwary archeologist, intending to date human activity by sampling charcoal from an ancient hearth, might be fooled into believing the site was remarkably ancient. Its apparent age would be over 43,000 years.

Other plant remains screened from the fill within Stanton's Cave are also notable. Hevly (this volume) reported quantities of juniper and cactus *(Opuntia)* seeds. Redbud pods *(Cercis)* and Utah agave appeared in certain pack rat middens: both grow outside Stanton's Cave at present. Detailed studies of the fossil flora of the ancient middens, some of which are located deep inside the cave, remains to be attempted. No obvious

Figure 12-3. North-South Trench in main room of Stanton's Cave looking north. The driftwood is below the figurine (arrow). Vertical scale is one meter long.

masses of juniper or piñon twigs, which commonly characterize fossil pack rat middens around Rampart Cave and in Hance Canyon, were seen in the Stanton's Cave middens. Robbins' preliminary study of cuticles in dung of Harrington's mountain goat revealed mainly grass, as one might expect of a grazer, as well as other herbaceous plants known to grow in the vicinity. Below the middle of the Stanton's Cave pollen profile (Robbins et al., this volume), *Artemisia* (probably Great Basin sagebrush) increases notably. Whether sagebrush itself occupied this part of the Inner Gorge at the time is less certain; however, *Artemisia tridentata* leaves were found by Phillips (1977) in Pleistocene middens near Rampart Cave, and shadscale, its common Great Basin associate, was found by both Phillips and Cole (1982). A fossil record of sage grouse *(Centricerus orophasianus)* from Stanton's Cave may be coeval. Had the vegetation of the upper elevations undergone a simple downward displacement in the last ice age, I would expect much more oak *(Quercus)* pollen below 30 cm, and much less *Artemisia*. Oak macrofossils were reported by Hevly (this volume) in the upper 20 cm only. While he found abundant juniper seeds between 20 and 30 cm, the remains of juniper were remarkably scarce below 30 cm where the *Artemisia* curve rises. The scarcity of both oak and piñon from the lower portion of the pollen record and the abundance of juniper only around 20-30 cm suggests that the vegetation of the type now growing 600-900 m higher in the Grand Canyon had not been displaced, rather it had withdrawn entirely (Cole 1982). My guess is that 15,000 years ago when mountain goats lived in Stanton's Cave, the plant community growing outside was dominated by sagebrush, shadscale, and occasional juniper (see also Cole 1982).

What was learned at Stanton's Cave? Sediments deposited on the floor are very shallow, less than one meter in depth, bottoming in driftwood, over 43,000 years. Four-thousand-year-old figurines were found on or just beneath the surface. Radiocarbon ages of 17,000 years were obtained on animal dung at a depth of only 50 cm. Fecal pellet preservation varied from excellent in the upper 50 cm to fair at greater depth.

Fossil pollen and seeds from the cave fill indicate major changes in vegetation, a change portended in the pioneering Grand Canyon pollen counts from Tséan Kaetan Cave by Roger Anderson (in Schwartz et al. 1958). During the last glaciation piñon and oak were scarce or absent in this portion of the Grand Canyon. Sagebrush and shadscale had invaded the Inner Gorge (Phillips 1977; Cole 1982). Some time much earlier, over 40,000 years ago, a mass of driftwood found its way into Stanton's Cave, perhaps reflecting some great flood, perhaps the surge of the river following collapse of a temporary dam upstream.

While prehistoric people built low rock walls on terraces near Stanton's Cave 900 years ago, they left little evidence of their activity in the cave itself. Much earlier, around 4000 years ago, the cave served a ceremonial or magical purpose to hunting-gathering people who left hundreds of twig figurines in the cave. The earliest suggestion of man's activity is enigmatic. The deposition of horn sheaths of Harrington's mountain goat and of its large pellets, along with bones of various extinct animals, ceased by about 11,000 years ago. This is the time of Clovis mammoth hunters elsewhere in Arizona. The possible linkage of these two events challenges further study of Southwestern cave infilling and content.

Thus, despite the shallow nature of their deposits, the caves of the Grand Canyon contain an unusual record of the nature of prehistoric life in arid America. Stanton's Cave yielded material of critical significance in testing current ideas about chronology and cause of large animal extinction. It yielded evidence, in the form of ancient driftwood, of river behavior different from any in recent times on the Colorado. Finally, from plant fossils in both cave earth and ancient wood rat nests Stanton's Cave yielded evidence that the "life zones" or vegetation pattern now familiar to Canyon visitors was not the same in the past. Over 12,000 years ago both vegetation and fauna were distinctively different from any of the associations found in the Grand Canyon today.

Through applying careful archaeological techniques and a paleoecological outlook to his excavation, Bob Euler has recovered some intriguing information on the late Pleistocene prehistory of the Grand Canyon. The importance of this and other Grand Canyon caves is clear. The caves and their contents are worthy of the best efforts of National Park Service managers to protect them. Carefully studied, the contents of the caves offer scientists a unique opportunity to test various competing theories on the nature of ice age climatic change, the extinction of late Pleistocene fauna, and the fate of prehistoric man himself. The contents of its caves, no less than of its scenery, are worthy of the adjective "spectacular."

References

Cole, Kenneth
1982 Late Quaternary zonation of vegetation in the eastern Grand Canyon. *Science* 172:1142-1145.

Houston, D.C.
1975 Ecological isolation of African scavenging birds. *Ardea* 63:55-64.

Mead, J.E.
1983 Harrington's extinct mountain goat *(Oreamnos harringtoni)* and its environment in the Grand Canyon, Arizona. Ph.D. dissertation, University of Arizona, 215 pages.

Murie, O.J.
1974 *A field guide to animal tracks,* 2nd edition. Boston: Houghton Mifflin Co., 375 pages.

Phillips, A.M.
1977 Pack rats, plants and the Pleistocene in the Lower Grand Canyon. Ph.D. dissertation, University of Arizona, 123 pages.

Schwartz, D.W., A.L. Lange and R. deSaussure
1958 Split-twig figurines in the Grand Canyon. *American Antiquity* 23:264-274.

Thompson, R.S., T.R. Van Devender, P.S. Martin, A. Long and T. Foppe
1980 Shasta ground sloth *(Nothrotheriops shastensis* Hoffstetter) at Shelter Cave, New Mexico: environment, diet and extinction. *Quatenary Research* 14:360-376.

Wilson, R.W.
1942 Preliminary study of the fauna of Rampart Cave, Arizona. Contribution to Paleontology, Publ. 530:169-185, Carnegie Institution of Washington.

Chapter 13

Conclusions

by

Robert C. Euler

Grand Canyon National Park

and

Department of Anthropology

Arizona State University, Tempe

Although Stanton's Cave had been entered by many individuals since Robert Stanton's first recorded visit in 1889, the data presented in this volume represent the first collected in a controlled scientific fashion.

The field excavations, supported by grants from the National Geographic Society, were designed to test the deposits in the cave. They were planned to preserve most of the fill of the cave floor for future scientific considerations.

With careful surface reconnaissance and highly controlled excavation of two test trenches aimed at as complete an ecological recovery as possible, a considerable body of new information, especially concerning the paleoenvironment of Grand Canyon, has been gathered. The various chapters of this volume present the analyses of archaeological, geological, and biological specimens observed and collected. They refer chronologically to the Archaic split-twig figurine complex dating from at most 4000 years before the present to the deposition of driftwood on the bedrock floor of the cave in excess of 43,000 years ago.

Unfortunately, no new information as to the culture of the makers of the figurines was recovered. To date no diagnostic artifacts have been found in association with the figurines that would provide definitive data as to the makers of these objects. The figurines described and depicted here, a greater number than any previously published, all conform essentially to known configurations.

Much of the fauna and flora from the lower deposits predate the Holocene and belong to the last Ice Age. The recovery of bones, horn sheaths, and dung (from which diet has been reconstructed) of the extinct mountain goat, *Oreamnos harringtoni,* when related to the more recent definitive study of that species in Grand Canyon (Mead 1983), is of considerable significance.

The pollen analyses indicate a change in environment of the area around the cave, and for Grand Canyon in general, apparently about the time of the Wisconsin pluvial. These show an alteration from a cold Great Basin desert condition to a warmer and wetter summer climate after about 11,000 B.P., a change better documented in recent pack rat midden records in the region (Cole 1982).

The avifauna recovered are also significant. A fossil robin, *Turdus grayii,* is the first such record. Numerous fragments of the large extinct vulture, *Teratornis merriami,* also were recovered. Bone radiocarbon dates of more than 15,000 years ago represent the first direct age determination for the species.

Considerable interest and research has been generated by hypotheses regarding how the driftwood was deposited in the cave. Two such possibilities have been presented here. One involves damming and ponding of the river; the other postulates a higher Pleistocene flow than any recorded historically. It seems to me to be in the best interests of science to present these two contrasting views. Ongoing research will undoubtedly provide the answers.

These are some of the data that make the Stanton's Cave deposits "spectacular" in the words of Paul Martin in the preceding chapter. He has made a good contribution relating the several floral and faunal taxa to each other and to their chronology.

The studies at Stanton's Cave are, however, but a beginning of paleoenvironmental investigations in Grand Canyon. If efforts by the National Park Service to prevent vandalism in the cave are successful, more biological and paleontological data await future investigators armed with better techniques than were available to me.

Even without future excavations, better correlations of the existing data can be made. The concluding comments that I made in a preliminary report on Stanton's Cave (Euler 1978:160) continue to be pertinent. I then remarked that:

"The biologists must look to the matter of process, and must pose and attempt to answer questions similar to those now being raised by archaeologists in other contexts (Plog 1974). To paraphrase those cultural queries, we need to wring from the Stanton's Cave data information, for example, as to why the environment changed as it did, why there is an apparent hiatus, why some change was slow and some rapid, and whether the data can be replicated elsewhere. These factors are probably due to certain variables and those which are important to paleoenvironmental change must be isolated. Only then will the excavated data from Stanton's Cave be put to its greatest use."

References

Cole, Kenneth
1982 "Late Quaternary zonation of vegetation in the Eastern Grand Canyon." *Science* 217:1142-1145.

Euler, Robert C.
1978 "Archeological and paleobiological studies at Stanton's Cave, Grand Canyon National Park, Arizona — A report of progress." *National Geographic Society Research Reports,* 1969 Projects. Washington: National Geographic Society.

Mead, Jim I.
1983 Harrington's extinct mountain goat *(Oreamnos harringtoni)* and its environment in the Grand Canyon, Arizona. Ph.D. dissertation. Tucson: University of Arizona.

Plog, Fred T.
1974 *The study of prehistoric change.* New York: Academic Press.